Karsten Berns | Ewald von Puttkamer

Autonomous Land Vehicles

Microsoft Dynamics NAV
by P. M. Diffenderfer and S. El-Assal

From Enterprise Architecture to IT Governance
by K. D. Niemann

ISSE/SECURE 2007 Securing Electronic Business Processes
by N. Pohlmann, H. Reimer and W. Schneider (Eds.)

Understanding MP3
by M. Ruckert

Process Modeling with ARIS®
by H. Seidlmeier

The New PL/I
by E. Sturm

www.viewegteubner.de

Karsten Berns | Ewald von Puttkamer

Autonomous Land Vehicles

Steps towards Service Robots

With 246 Illustrations, 4 Tables and 16 Algorithms

VIEWEG+
TEUBNER

Bibliographic information published by the Deutsche Nationalbibliothek
The Deutsche Nationalbibliothek lists this publication in the Deutsche Nationalbibliografie;
detailed bibliographic data are available in the Internet at http://dnb.d-nb.de.

1st Edition 2009

Editorial Office: Sybille Thelen | Maren Mithöfer

Vieweg+Teubner is part of the specialist publishing group Springer Science+Business Media.
www.viewegteubner.de

Cover design: KünkelLopka Medienentwicklung, Heidelberg
Printing company: Krips b.v., Meppel
Printed on acid-free paper
Printed in the Netherlands

ISBN 978-3-8348-0421-1

Preface

This textbook results from a series of lectures concerning autonomous mobile robots which have been held at the University of Kaiserslautern between 1999 and 2009. Methods and algorithms are introduced which can be used for developing complex autonomous land vehicles. Starting from historical remarks and application areas of service robots, the vehicle kinematics modeling is introduced and examples of the drive kinematics of different vehicles are given. Thereafter, typical sensors and sensor systems are described which are used to determine the internal state of the machine and its operational environment. Localization, i.e the determination where the robot is, is still a difficult problem. In the textbook, several methods are discussed which can be used under specific preconditions. Map building as well as navigation strategies complement the set of basic methods. The last two chapters deal with the questions of how to compile the above mentioned methods using powerful control architecture and what frameworks to use to support the development process.

This textbook is written for beginners and advanced students from the fields of computer science, mechanical engineering, and electrical engineering, specializing in autonomous mobile systems. The book is also suited for engineers with a special interest in the development of wheel driven service robots.

The writing of the manuscript was only possible with the assistance of several researchers of our Robotics Research Lab. Special thanks to Sebastian Blank, Tim Braun, and Martin Proetzsch for proof-reading and editing. Helpful contribution has been given by Daniel Schmidt (chapter kinematics), Carsten Hillenbrand and Sebastian Prehm (chapter sensors), Jan Koch, Bernd-Helge Schfer, and Norbert Schmitz (chapter localization), Christopher Armbrust, Tim Braun, and Jens Wettach (chapter mapping), Tobias Fhst (chapter SLAM), Tobias Fhst and Jens Wettach (chapter navigation), Christopher Armbrust and Martin Proetzsch (chapter control architecture), Max Reichardt (chapter frameworks).

Karsten Berns   Ewald von Puttkamer

Contents

1 Introduction

Industrial robots, which are among the most important elements for industrial automation, are the biggest commercial market for robotics. Since the 1970s more than one million units are in use in fields like car manufacturing or the chemical industry. These robots are operating in highly structured environments. The tasks performed by this type of robots are monotone and restricted due to a low flexibility. An economic growth of the robotics industry will only be achieved if the systems become mobile, adaptive to a dynamic environment and can be used for different tasks. The robots mentioned above belong to the class of service robots. A service robot can be defined as a system which operates semi- or fully autonomously to perform services useful to the well-being of humans and equipment, excluding manufacturing operations.[1]

Starting from autonomous guided vehicles, sewer inspection robots, cleaning robots or systems for entertainment, the number of service robots has increased tremendously during the last 15 years. It is expected that the service robot market will be growing by a factor of eight to about 50 billion USD in 2025. Most of these service robots belong to the class of autonomous mobile robots (AMR).

This book is an introduction to basic techniques and methods which allow the development of such machines. In the following, a short overview of the problem areas of AMRs, their applications and a historical survey are given.

1.1 Autonomous mobile robots

Autonomous mobile robots (AMR) can be defined as robots able to navigate through the environment in an autonomous way while performing goal-oriented tasks. They can be classified according to their operational environment into unmanned ground vehicles (UGV), unmanned water vehicles (UWVs), autonomous underwater vehicles (AUVs) and unmanned aerial vehicles (UAV). They can be used for different service tasks like autonomous inspection, surveillance or maintenance. Figure 1.1 demonstrates typical

---

[1] World Robotics 2003, United Nations and International Federation of Robotics

AMRs. The insect-type robot LAURON [GB01] developed at FZI, Karlsruhe (Germany) is designed for navigation in rough terrain. The autonomous helicopter H3 of the TU Berlin (Germany)[2] was built for search and transportation tasks while the RoboTuna project of MIT, Boston (USA)[3] examines biologically inspired underwater locomotion.

Figure 1.1 LAURON III (left), UAV H3 (middle), and RoboTuna II (right)

This textbook focuses on UGVs that make use of wheels for locomotion purposes. Most methods concerning localization, mapping and navigation can also be transferred to other types of AMRs. Common requirements for AMRs are autonomy and autarchy. Autonomy means that the system can decide self-dependently. One can separate between fully autonomous, in which the system decides totally by itself, and semi-autonomous, in which some decisions are made by a human operator. The decisions are normally based on incomplete knowledge about the environment and might be wrong considering the global task. The term autarchy signifies that the energy supply is carried along on the robot. It is clear that these are requirements essential for tasks in which a mobile robot is necessary.

To fulfill the requirements for autonomy of an AMR, the following features are essential:

Mobility This term describes the ability of the robot to move to specific positions in the operational environment. These positions could be in the local surrounding but also far away.

Adaptivity to unknown situations Because of a high dynamic of the environment the AMR will be confronted with situations which have not been specified before. Therefore, adaptivity is a key feature for AMRs.

Perception of environment For navigation and the ability to fulfill application tasks it is essential to retrieve information of the environment

---

[2] http://pdv.cs.tu-berlin.de/lfafr/index.html
[3] http://web.mit.edu/towtank/www/Tuna/Tuna2/tuna2.html

around the vehicle. The main problems arise from incomplete and noisy basic data.

Knowledge acquisition Because a complete model of the operational environment of the robot cannot be described a priori, the AMR must have the ability to acquire new knowledge while operating.

Interaction ability AMRs must also be able to get commands and new tasks from an operator. Very often speech and gesture are more suitable than standard techniques.

Safety The AMR must not destroy itself, damage any objects or hurt humans. An emergency stop is the simplest technique and should be avoided by using a prediction of critical situations.

Realtime processing Computer and software architecture able to deal with hard real-time requirements.

To implement the above mentioned features, different problem areas have to be handled. Given a mobile robotic platform, the first step towards a solution to a service robot problem is the modelling of its kinematics and dynamics. This includes e. g. the relation between wheel velocities and the robot movements or the influence of wheel slippage. Based on these models, simple navigation tasks can be described that do not consider obstacles in the environment. To avoid collisions, sensor systems are needed to detect different types of obstacles like stairs, furniture or vegetation. Based on the measurement principle, the extracted information is often noisy and incomplete. This makes it necessary to use different sensors and fusion algorithms for the measured values. The information is used by the collision avoidance strategies to decide whether an emergency stop or an evasion of obstacles is the best solution.

If the robot is supposed to be able to drive along a predefined path given as intermediate positions and orientations according to a fixed frame, it has to know where it is. This problem is called the localization problem. Because of disturbances from the environment (e. g. slippage of the wheel) it is not possible to solve this problem using solely kinematic or dynamic models in a precise way. Therefore, additional methods have to be taken into account.

For the execution of navigation tasks it is often helpful to describe the operational environment with maps. Therefore, one has to decide which features are extractable from the environment, and how to represent them. These maps can be used for navigation. This includes path planning under given quality criteria like time-optimum or length of the travelled path. Depending on the application, the AMR must recognize whole scenes or only specific objects that are to be manipulated. Problems are the extraction of features which lead to an unique identification and dynamic changes of the situation.

In the following, methods and techniques for the solution of the described problems will be introduced.

1.2   Applications of autonomous mobile robots

In general, all robotics applications can be classified according to the degree of unstructuredness of the environment and the degree of autonomy that is necessary for executing the task (see figure 1.2). Industrial robots are normally used in a well defined structured environment. For example, welding robots in car manufacturing get the exact position and orientation of where the segments of the chassis have to be welded together and where these parts are located. Therefore, the number of sensors that have to observe the manufacturing process is very low. The task execution is restricted to a fixed set of commands, which cannot cope with unforeseen conditions. Thus, the degree of autonomy of the task execution is also low.

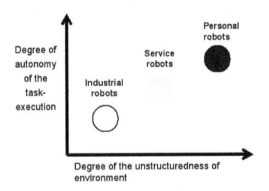

Figure 1.2   Classification of AMR systems with respect to the field of application

In contrast to that service robots have to operate within more unstructured environments. It is not possible to depict these environments completely. Consider for example an underwater robot that is intended to inspect a pipeline. In this case a model of the environment and its dynamic conditions like critical flows could not be given a priori. Furthermore, the use of specific sensor systems like camera systems will deliver a limited model of the environment because of e. g. the water turbidity. These disturbances cause a strong need to adapt to the current situation for he task execution.

Personal robots on the other hand are designed to be general purpose machines suitable for performing a huge amount of different tasks in arbitrary environments. A humanoid robot, for example, can be applied as a servant in a household. It should be able to clean dishes and windows or wash clothes. The operational environment is located both inside the house and in the garden. In the following only application areas and some examples in which specific service robots are used will be presented.

Service robots can be classified based on their application areas. A detailed description of these systems and its applications can be found in [SS04, Ich05].

Transportation Typical transportation systems are industrial automated guided vehicle (AGV), transportion systems for goods or support systems for handicapped persons. Because these vehicles operate mainly in the proximity of humans, the safety requirements are very high. The main focus lies on path planning under different environmental conditions and taking obstacles into account. A typical example of an AGV is TransCar of the company Telelift, Puchheim (Germany).[4] It moves routine and on-demand deliveries of medical supplies across multiple-floor facilities. Another transportation system is the autonomous truck Actros of the company UZIN, Ulm (Germany)[5], which carries production materials on the factory premises. The wheelchair Rolland of the University of Bremen (Germany)[6] is developed for the transport of elderly and handicapped people.

Surveillance Surveillance robots have the task of monitoring buildings and areas both in- and outdoor. Normally, a fixed path or waypoints are given, which are frequently visited. Besides the implementation of adequate navigation strategies, the detection of irregularities in structured and unstructured terrain has to be solved. In addition, the system has to distinguish between intruders and authorized persons. The robots Mosro and the Ofro of Robowatch, Berlin (Germany)[7] are representatives for this application.

Exploration Several robot systems for the application in environments that are either hazardous or non-accessible for humans were developed in the last years. These systems have to be immune to any disturbances

---

[4] http://www.swisslog.com/index/hcs-index/hcs-systems/hcs-agv/
hcs-agv-transcar.htm

[5] http://www.goetting.de/de/multimedia/videos/fox_auf_vox.flv.html

[6] http://www.informatik.uni-bremen.de/rolland/

[7] http://www.robowatch.de/index.php

and unforeseen situations. They must be implemented in a way that an interaction of a human operator with the system via telecontrol is possible. The planetary rovers Spirit and Opportunity of the National Aeronautics and Space Administration (NASA, USA) landed on Mars in 2004 and are still operating.[8] The ROBOVOLC vehicle developed in the course of an EU project led by the University of Catania (Italy)[9] was used to explore the Etna volcano.

Inspection and maintenance Inspection and maintenance tasks represent one of the biggest application areas for service robots. They are used to analyze plants, buildings or large technical devices like ships. They can also be employed to clean or repair them. The operational environment could be subjected to extreme conditions like high and low temperatures or any kind of liquids. Besides methods for inspection and maintenance, the exact positioning is a great challenge. In sewer inspection, for instance, robots are used to detect broken pipes or other damages, which may lead to ground contamination. Kairo of the FZI, Karlsruhe (Germany) is a snake-like robot for the autonomous inspection of pipe networks [SKBD01]. The Robair system of the London South Bank University was developed to inspect rows of rivets for loose ones and cracks at the wings and fuselage of airplanes [SPCB06]. Other applications are underwater, like inspection and repairing of pipelines. The teleoperated vehicle Spider was designed by the company Cybernetix, Marseille (France) [SHW04] and visually analyses pipelines up to a depth of 1500 m.

Harvesting For forestry and agriculture, different service robots have been developed to reduce costs and to save resources. One problem area arises from the motion in uneven and unstructured terrain; another from the detection of the crop. The six-legged robot Harvester developed by the company Plustech, a Finnish John Deere subsidiary, is used for cutting trees in rough forests [SS00]. An autonomous fruit picking machine (AFPM) is a further example used for apple harvesting [BDB+07].

Housekeeping Housekeeping robots are an old dream presented in several science fiction stories. In the last years, several robots have been designed to vacuum-clean, to clean windows or to support people with their housework. Beside complex manipulation and navigation tasks,

---

[8] http://marsrovers.nasa.gov/home/
[9] http://www.robovolc.dees.unict.it/activity/activity.htm

these robot systems often have to interact with human operators verbally or based on gestures and mimic. Trilobite of the company Electrolux was one of the first vacuum cleaner products.[10] Equipped with ultrasonic sensors, Trilobite is able to avoid obstacles. When the battery load runs low, the robot drives back to the charging station. More sophisticated research robots for housekeeping are the humanoid robots ARMAR of the University of Karlsruhe (Germany)[11] and Care-O-bot[12] of the Fraunhofer Institute IPA, Stuttgart (Germany). Both machines are able to perform manipulation tasks, like dish washing, or fetch and carry operations.

Edutainment Edutainment robots combine education and entertainment. The idea to motivate learning with the help of interesting robot systems can be found on all education levels. For pupils, Lego Mindstorms[13] is often used to introduce them to mechatronics as well as programming robotic systems. The RoboCup competition[14] inspires students worldwide to delve into robotics methods and technologies. Research areas are multi-agent systems, object tracking and game strategies. Other projects for edutainment are museum guides like TOURBOT.[15] This robot can be used as an interactive tour guide, providing individual access to museums' exhibits and cultural heritage over the Internet, or as a flexible, on-site museum guide.

1.3 Historical overview

Among the first AMRs to be mentioned in the literature is a machine called ELSIE (Electro-light-sensitive Internal-External)[16], designed by W. Grey Walter in the 1940s and 50s in England. It is a rather simple device equipped with very basic sensors and actors in order to enable ELSIE to localize a light source in its environment and approach the source's position. A simple collision avoidance mechanism was available. ELSIE could be regarded as the first autonomous mobile robot able to react by itself to specific stimuli of the operation environment. The control was based on analog circuits.

---

[10] http://trilobite.electrolux.de/
[11] http://www.sfb588.uni-karlsruhe.de/
[12] http://www.care-o-bot.de/
[13] http://mindstorms.lego.com/
[14] http://www.robocup.org/
[15] http://www.ics.forth.gr/tourbot/
[16] See http://www.extremenxt.com/walter.htm

It took until the late 1960s for the first more serious AMR to be developed. The new robot developed by the Stanford Research Institute (SRI) was named Shakey, see figure 1.3.[17] It was equipped with a TV camera, a triangulating range finder, and bump sensors and made use of programs for perception, world-modeling, and acting. Due to the increased need for computation performance, it consisted of both an on-board and an off-board computer (PDP-10) connected via a radio link. Vision and planning were performed off-board, while all other functions were performed by the on-board unit. The revolutionary new approach introduced with this platform, however, was hierarchical control, still a common principle used in modern robots.

Figure 1.3 Shakey, the forefather of autonomous mobile indoor robots

---

[17] http://www.ai.sri.com/shakey/images.php

Shakey's control architecture consists of three levels: Low-level routines take care of simple tasks like moving, turning, or route planning. Intermediate-level routines combine the low-levels ones in order to be able to perform more complex tasks. On the highest hierarchy level routines to make and execute plans are found. Shakey can therefore be seen as the first cognitive robot which was able to solve complex task planning problems. Shakey had the task, for example, to move an object located on a platform in its operational environment. Because it was unable to move on the platform, Shakey first plans how to reach it. The solution was to move a ramp to the platform first and then drive over the ramp to the object. Therefore, the general problem solver STRIPS was used. The control programs are implemented with the programming languages Fortran and Lisp.

A few years later, in the early 1970s, the NASA in cooperation with Jet Propulsion Laboratory (JPL) at Pasadena (USA), launched a program intended to provide real-time control, reduce support requirements, and enhance performance and reliability. One result of this program was the Mars rover. Unfortunately, its degree of autonomy wasn't significantly higher than the one of other platforms developed in those early days of AMR research. However, it was equipped with a modified Stanford arm as manipulator and a variety of sensors like laser range-finders, or stereo TV cameras, as well as tactile and proximity sensors.

Its navigation system was based on a gyroscopic compass and optical wheel encoders employed for dead-reckoning. Again the need for the computational performance made an off-board computer inevitable as an addition to the 32 K memory on-board system to construct a "world model" and perform planning. The robot's basic ability was to analyze a simple environment with a limited number of objects, plan a path and follow it to a goal.

The representation of the world used for this purpose was a segmented terrain model consisting of map sectors of reasonable size. Each sector was assigned an attribute representing the accessibility for the robot: Hence, a certain sector was either not traversable or unknown. All other sectors were assumed to be traversable.

The Stanford Cart was developed by Hans Moravec of AI Lab, Stanford (1973–1981). This semi-autonomous controlled robot was equipped with a stereo-camera system, in order to generate 3D images. Since image processing on board was not possible, the image was sent to an external computer. After the image was processed, the distance information of the objects was sent back. The objects were described in polar-coordinates.

For the determination of an optimal path, a tree-search approach was used. The CMU mobile Robot (1981–1984) continued the Moravec research. This cylindrically shaped vehicle with a height of 1 m and a diameter of 30 cm had 12 on-board processors. As control architecture, a three level task-based approach was used, consiting of a planner, initiator, sensor processing and the motor control. Until now, several mobile robots apply such a control architecture.

Meldog was a remarkable approach to develop a autonomous robot. Meldog, which was built as a kind of motor cycle, was developed at the Mechanical Engineering Lab in Japan (1979–1983). It was a mobility aide to blind people and acted as a kind of robotic seeing-eye dog. The vehicle was able to detect obstacles in the nearby environment with ultrasonic sensors. Based on this information collision avoidance was performed. Furthermore, the speed adjustment to keep a distance of 1 m to the operator was triggered by ultrasonic measurements. With the help of a wired link, the operator was able to control the machine (left, right, straight, stop). Maps were also used to support path planning.

In Europe the development of AMRs was pushed forward by EUROMICRO, organizing micromouse contests from 1980 on at their annual meetings. The competition was to let a vehicle find the middle of a maze, made up of small walls of 10 mm thickness and 50 mm height in tiles of 17 cm side length. In 1981 a first Micromouse, as sketched in figure 1.4 [GHvP81], with a diameter of 15 cm, equipped with two coupled microprocessors, was brought to the meeting at Kopenhagen by the researchers of the University of Kaiserslautern. One of the microcontrollers was used to control the vehicle, the other one to solve the maze. Afterwards, a new design was developed for a much smaller vehicle. This was the micro mouse Speedy Gonzales (dimensions 130 mm×100 mm) [HK88] which managed to solve the maze successfully. Figure 1.5 shows a sketch of the robot and a picture of the vehicle in a maze.

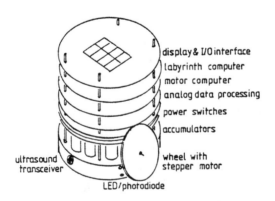

display & I/O interface
labyrinth computer
motor computer
analog data processing
power switches
accumulators
ultrasound transceiver
wheel with stepper motor
LED/photodiode

Figure 1.4 Concept of the Micromouse

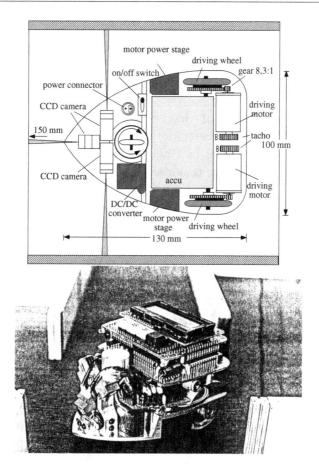

Figure 1.5 Micro mouse Speedy Gonzales

Besides this, several other activities were started in Germany in the 1980s. Several groups in Munich, Darmstadt, Berlin and Aachen developed mobile robots. At the TH Karlsruhe Prof. Rembold and his group started the KAMRO project (Karlsruhe Autonomous Mobile Robot, 1985–1995). The autonomous mobile robot KAMRO (see figure 1.6) was one of the first robots to perform assembly tasks autonomously. It was equipped with two arms (Puma 260 manipulators) with a gripper and an integrated camera, a mobile platform with an omnidirectional drive system, and different sensors for navigation, docking and manipulation. The tasks could be described on different levels: assembly precedence graphs, implicit elementary operations (pick, place) and explicit elementary operations (grasp, transfer, fine motion, join, exchange, etc.). KAMRO provided the basis of a whole series of autonomous vehicles at the University of Kaiserslautern.

In the 1990s, a lot of wheel-driven robots have been developed all over the world. Several companies started to build wheel-driven robots which were mainly sold to the research market. Besides this, service robots like vacuum cleaners, sewer inspection robots or autonomous transport vehicles have been used.

From the research point of view, Sojourner (see figure 1.7), developed at the JPL (1994–1997), was certainly one of the robotic highlights of the 1990s. This small robot driven by 6 wheels was the first machine on Mars. The machine was controlled by teleoperation from Earth. Path planning, for example, was done with the help of a simulation system. When an optimal path was selected, its parameters were transferred to Sojourner on Mars. Obstacle detection and avoidance was directly performed on the machine. Sojourner was equipped with a gripper able to collect samples. A first analysis of the collected material was also done on the machine.

Figure 1.6 KAMRO of the University of Karlsruhe (Germany) designed for a flexible production (courtesy of Prof. Dillmann, TH Karlsruhe)

Figure 1.7 The Mars exploration rover Sojourner developed by NASA and JPL (from http://www-robotics.jpl. nasa.gov/roboticImages/ img811-67-browse.jpg)

This short historical survey is certainly not complete. In the 1970s and 80s there have been further autonomous mobile robot research projects mainly in the US, Japan and Europe. During this period, the first AGV were used in factories to automatize production. Starting in the 1990s, service robots have been developed for all the above mentioned application areas. Most of the systems did not get beyond a prototype stage. Besides technical problems, the high cost was the main reason that these robot systems have not been placed on the market.

1.4 Book overview

After the above overview of the AMR topics, problems and applications, a short historical survey is given. Chapter 2 focuses on the kinematics of wheel-driven AMRs. Based on the rotational speed of typical wheel types, a kinematic model for the computation of the vehicle's linear and angular velocities is derived. For specific drive systems it is easier to apply a geometrical solution to the kinematic problem. Some examples of this are given.

The kinematic model is a foundation for several topics presented in the book. The description of the robot's state and the detection of objects in its environment are essential for tasks like localization and navigation. In chapter 3, the most important sensors and sensor systems for this purpose are introduced. Starting from tactile and pose[18] measurement sensors, different types of ranging and vision sensors are presented. Distance sensors are often the main source of information because they can be used in manifold ways e. g. for safety, 3D reconstruction or collision avoidance.

Despite sophisticated sensor systems like GPS, precise localization is still an open research problem. In chapter 4, the problem of localization is introduced and different techniques are presented. This includes dead-reckoning, localization based on optical flow, feature based methods and approaches using landmarks.

Complex AMR applications need an adequate map for the representation of the environment and for planning. In chapter 5, a classification of the different map types is given. Later, the concept for building different types of maps like metrical, grid, sector, topological and hybrid maps is discussed.

A precise determination of the position and orientation of the robot is required for map building. On the other hand, if a map exists and features of the environment can be correlated with it, one can derive the robot pose. SLAM methods generate new maps and estimate the pose of the robot in a simultaneous way. Chapter 6 gives a short overview of SLAM and presents solutions to some subproblems like the merging of maps.

---

[18] The pose describes the position and the orientation.

Strategies for moving from one location to another one in known or unknown environments are summarized under the term navigation. In chapter 7, algorithms for local and global path planning as well as path control are described. Local path planning covers both algorithms for planning based on different map types and methods of basic navigation abilities like collision avoidance.

The structuring of all control components according to functional and non functional requirements is done using control architecture. Chapter 8 introduces different standard architectures for AMRs. The main focus of the chapter is behavior based control architectures. The iB2C architecture is presented as an example and applied to specific control problems.

The book ends with an survey of software frameworks for robotic applications (chapter 9). Robotic frameworks support the development process. The implementation of control algorithms for complex AMRs is not possible without a suitable framework. In this chapter different features of frameworks are discussed and a comparison of well known examples from literature is given.

2 Kinematics

In this chapter, the foundation for path planning and navigation of a wheel-drive robot is given. First the basic formulas, which allow to describe the motion of the vehicle in a 3D environment, are introduced. Then the solution to the kinematics problem considering different types of wheels fixed at specific positions on the vehicle platform is described. At the end of the chapter, geometric kinematics solutions for typical wheel-driven robotic platforms are presented.

2.1 Basics

In the following section, a short introduction to the kinematics calculation (pose and velocity) is given and applied to a standard wheel-driven robot. Figure 2.1 shows two successive positions of a vehicle in a 3D environment. The pose of the robot is represented by a frame through the kinematic center of the robot (see figure 2.2). Path planning can be reduced to finding a navigation strategy which transforms the starting frame to the next frame. In a 2D scenario (e. g. driving in an office environment) a transfer vector (x,y,α) is determined which describes the displacement in x- and y-direction as well as the orientation represented as rotation around the z-axis. In the next section it is shown how – based on the rotational speed of the vehicle wheels – the velocity of the kinematic center can be calculated. Through the integration of the velocity vector over time, the pose change is determined.

In general the pose of an arbitrary object in the Cartesian space can be described as a six-tuple $(x,y,z,\alpha,\beta,\gamma)$. The position vector $^O\vec{u}$ in an object frame O can be presented in base frame coordinates $^B\vec{u}$ by

$$^B\vec{u} = \begin{pmatrix} x \\ y \\ z \end{pmatrix} + {}^B_O R(\alpha,\beta,\gamma){}^O\vec{u} \tag{2.1}$$

with $(x,y,z)^T$ being the translational vector between the origin of the two frames and R the respective (combined) rotation matrix.

Figure 2.1 Transformation of robot coordinate systems

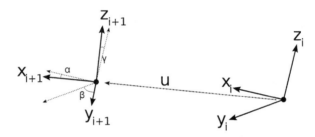

Figure 2.2 Schematics depicting the coordinate systems in figure 2.1

Another possibility to express the same as above are homogeneous transformation matrices. Those 4×4 matrices (for 3D space) are composed as shown below:

$$\left[\begin{array}{cc} R & u \\ P & s \end{array} \right] \tag{2.2}$$

with

R: 3×3 rotation matrix

u: position vector $u = (u_x, u_y, u_z)^T$

P: perspective transformation (in general $P = (0,0,0)$)

s: scaling factor (in general $s = 1$)

Matrix R usually is a combination of several rotations around the elementary axis. The rotation matrix $R_x(\alpha)$ describes a rotation around the x-axis of an arbitrary coordinate system via the angle α. The other two required matrices are defined the same way.

$$R_z(\alpha) = \begin{bmatrix} \cos\alpha & -\sin\alpha & 0 \\ \sin\alpha & \cos\alpha & 0 \\ 0 & 0 & 1 \end{bmatrix} \tag{2.3}$$

$$R_x(\alpha) = \begin{bmatrix} 1 & 0 & 0 \\ 0 & \cos\alpha & -\sin\alpha \\ 0 & \sin\alpha & \cos\alpha \end{bmatrix} \tag{2.4}$$

$$R_y(\alpha) = \begin{bmatrix} \cos\alpha & 0 & \sin\alpha \\ 0 & 1 & 0 \\ -\sin\alpha & 0 & \cos\alpha \end{bmatrix} \tag{2.5}$$

There are two possible ways of expressing linked rotations: The roll, pitch, yaw system and Euler angles. The former implies the linked rotations are performed around the fixed axes of the coordinate system while the latter uses variable rotation axes. A linked rotation around, for instance, the x-axis (angle α) then x-axis (angle β) and finally z-axis (angle γ) can be expressed as:

$$\text{Euler:} \qquad R_s = R_z(\alpha) \cdot R_x(\beta) \cdot R_{z'}(\gamma) \tag{2.6}$$
$$\text{Roll-pitch-yaw:} \qquad R_s = R_z(\alpha) \cdot R_y(\beta) \cdot R_x(\gamma) \tag{2.7}$$

As one can see, the axes in the latter case remain fixed while in the first case rotations along "new" axes are performed. In case of terrestrial robotics the roll, pitch, yaw system is selected to transfer the frames into each other. An illustration of the concept can be found in figure 2.3.

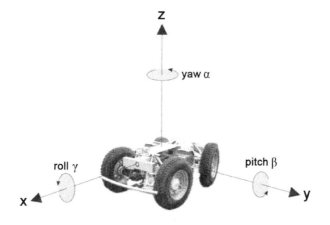

Figure 2.3 Roll, pitch, yaw: rotation along fixed coordinate axes

The system originates from the field of aviation. The difference to terrestrial robotics is the yaw-axis, which is changed from facing down to up for usability reasons.

Since an arbitrary orientation in 3D space can be achieved using only three rotations, the resulting linked rotation R_s can be expressed as presented in equation 2.8. For writing and reading convenience the abbreviations $s\alpha$ for $\sin(\alpha)$ and respectively $c\alpha$ for $\cos(\alpha)$ will be used in the following equations:

$$R_s = \begin{bmatrix} c\alpha c\beta & c\alpha s\beta s\gamma - s\alpha c\gamma & c\alpha s\beta c\gamma + s\alpha s\gamma \\ s\alpha c\beta & s\alpha s\beta s\gamma + c\alpha c\gamma & s\alpha s\beta c\gamma - c\alpha s\gamma \\ -s\beta & c\beta s\gamma & c\beta c\gamma \end{bmatrix} \tag{2.8}$$

The velocity vectors can be calculated similar to the transformation of the pose from one frame to another. Suppose there is a linear velocity vector $^B\vec{v}_Q$ of an arbitrary point Q presented in frame B which should be transformed to frame A. This transformation can be calculated as

$$^A\vec{v}_Q = {}^A_B R\, {}^B\vec{v}_Q \tag{2.9}$$

If the origin of frame B has also a linear velocity relative to frame A then

$$^A\vec{v}_Q = {}^A\vec{v}_{OB} + {}^A_B R\, {}^B\vec{v}_Q \tag{2.10}$$

If in addition point Q is rotating around an arbitrary axis with the rotational velocity $^A\Omega_B$ then the linear velocity can be calculated with

$$^A\vec{v}_Q = {}^A\vec{v}_{OB} + {}^A_B R\, {}^B\vec{v}_Q + {}^A\Omega_B \times {}^A_B R\, {}^B Q \tag{2.11}$$

A rotational vector $^B\vec{\omega}$ related to frame B can be transferred to frame A with

$$^A\vec{\omega} = {}^A_B R\, {}^B\vec{\omega} \tag{2.12}$$

If there are several segments which are connected to each other by rotational joints or prismatic joints, the rotational and the linear velocity can be calculated step by step for each segment starting from the base frame. The rotational velocity $^{i+1}\omega_{i+1}$ and the linear velocity $^{i+1}\vec{v}_{i+1}$ due to frame $i+1$ can be determined as:

$$^{i+1}\omega_{i+1} = {}^{i+1}_i R \cdot {}^i\omega_i + \dot{\theta}_{i+1}\, {}^{i+1}e_{z_{i+1}} \tag{2.13}$$

$$^{i+1}\vec{v}_{i+1} = {}^{i+1}_i R \left({}^i\vec{v}_i + {}^i\omega_i \times {}^i P_{i+1} \right) \tag{2.14}$$

with ${}^i P_{i+1}$ the vector in direction of the segment, i, ω_i the rotational velocity and θ_i the rotation of segment i around the elementary z-axis. It has
to be taken into consideration that ${}^{i+1}_i R$ is the inverse of the orientation
transformation ${}^i_{i+1} R$ from frame i to $i+1$. Because ${}^i_{i+1} R$ is an orthogonal
matrix its inverse ${}^i_{i+1} R^{-1} = {}^{i+1}_i R$ is just the transposed matrix ${}^i_{i+1} R^T$ (see
equation 2.15).

$$
{}^{i+1}_i R(\theta) = {}^i_{i+1} R^{-1}(\theta) = {}^i_{i+1} R^T(\theta) = \begin{pmatrix} c\alpha & s\alpha & x_w \\ -s\alpha & c\alpha & y_w \\ 0 & 0 & 1 \end{pmatrix} \tag{2.15}
$$

2.2 Wheel kinematics

In the following, the stepwise calculation of the linear and the rotational
velocity will be applied to wheeled vehicles operating in a 2D environment.
The question to be answered is how the rotational velocity of each wheel can
be determined, as the kinematic center is moved with a linear velocity $\dot{x}$ due
to the x-axis, $\dot{y}$ due to the y-axis and $\dot{\theta}$ the rotational velocity around the
z-axis. Also the inverse of this problem should be determined. Therefore, the
basic wheel types shown in figure 2.4 will be considered.

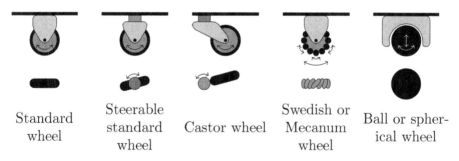

Standard wheel   Steerable standard wheel   Castor wheel   Swedish or Mecanum wheel   Ball or spherical wheel

Figure 2.4 The basic wheel-types

It is obvious that for the standard and steerable standard wheel there
should be no sliding orthogonal to the wheel plane (the velocity must be
zero). The linear speed of each of these wheels in rolling direction can be
calculated with $r\dot{\psi}$ (r is the radius of the wheel and $\dot{\psi}$ is the rotational speed
of the wheel). This formula could also be applied for the rolling speed of the
castor wheel and the spherical wheel. If d is the offset between the wheel axis
and the vertical axis of rotation, the linear velocity orthogonal to the wheel

plane is the rotational velocity around the vertical axis times the length of the offset $(-d_c\dot{\beta})$.

In case of the Swedish or Mecanum wheel, in which passive rollers are mounted in an angle γ (γ is normally 45° or 90°) on the perimeter of the main wheel, the linear velocity in rolling direction is $r\dot{\psi}\cos\gamma$. Because the Swedish or Mecanum wheel is able to move in an omnidirectional way, the speed orthogonal to the wheel plane can be calculated as $r\dot{\psi}\sin\gamma + r_{pr}\dot{\psi}_{pr}$ (with r_{pr} the radius and $\dot{\psi}_{pr}$ the rotational speed of the roller).

To determine the velocity of a wheel due to the velocity vector $\vec{v} = (\dot{x},\dot{y},\dot{\theta})^T$ of the kinematic center, one must first define the robot coordinate frame, which has its origin in the kinematic center of the vehicle (see figure 2.5). By definition, $\alpha = 0$ if the normal vector of the wheel plane is located on the x-axis and has the same orientation. β determines the angle between the straight line through the kinematic center and the fixing point of the wheel and the y-axis of the wheel frame. Parameter d is the distance from the kinematic center to the fixing point of the wheel on the chassis. The wheel coordinate system has its x-axis in the rolling direction and the y-axis as the normal to the wheel plane. The linear speed of the wheel is in the direction of the x-axis of the wheel frame. This parameter definition can be used for all wheel types. In case of the Swedish or Mecanum wheel an additional parameter γ has to be introduced, which describes the angle between the x-axis of the wheel and rolling axis of the rollers.

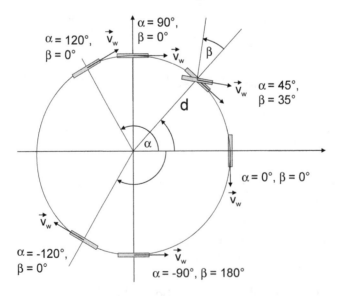

Figure 2.5 The parameters for solving the kinematic problem for different wheel positions

Supposing the velocity vector $\vec{v} = (\dot{x},\dot{y},\dot{\theta})^T$ of the kinematic center is given, equation 2.13 and equation 2.14 should be applied to calculate the linear velocity of a standard wheel. Thus, we receive $^1\omega_1 = (0,0,\dot{\theta})^T$ and $^1(\vec{v})_1 = (\dot{x},\dot{y},0)^T$. Because there is no additional rotational speed, all $^i\omega_i = (0,0,\dot{\theta})^T$ (see equation 2.13). A stepwise application of equation 2.14 will deliver the following $^i(\vec{v})_i$.

After the rotation around the z-axis with angle α, $^2\vec{v}_2$ is calculated as:

$$^2\vec{v}_2 = \begin{pmatrix} c\alpha\dot{x} + s\alpha\dot{y} \\ -s\alpha\dot{x} + c\alpha\dot{y} \\ 0 \end{pmatrix} \qquad (2.16)$$

Due to the translation d, $^3\vec{v}_3$ is:

$$^3\vec{v}_3 = \begin{pmatrix} c\alpha\dot{x} + s\alpha\dot{y} \\ -s\alpha\dot{x} + c\alpha\dot{y} + d\dot{\theta} \\ 0 \end{pmatrix} \qquad (2.17)$$

The last rotation around the z-axis with angle $\beta - 90°$ transfers the x-axis of the last frame to the rolling direction of the wheel. For the calculation of $^4\vec{v}_4$ in the next equation, $\sin(\beta - 90°) = -\cos(\beta)$ and $\cos(\beta - 90°) = \sin(\beta)$ is used.

$$^4\vec{v}_4 = \begin{pmatrix} s(\alpha + \beta)\dot{x} - c(\alpha + \beta)\dot{y} - c\beta d\dot{\theta} \\ c(\alpha + \beta)\dot{x} + s(\alpha + \beta)\dot{y} + s\beta d\dot{\theta} \\ 0 \end{pmatrix} \qquad (2.18)$$

The last step is to equalize the linear velocity vector of the standard wheel to that of equation 2.18.

$$\begin{pmatrix} s(\alpha + \beta) & -c(\alpha + \beta) & -c\beta d \\ c(\alpha + \beta) & s(\alpha + \beta) & s\beta d \\ 0 & 0 & 0 \end{pmatrix} \begin{pmatrix} \dot{x} \\ \dot{y} \\ \dot{\theta} \end{pmatrix} = \begin{pmatrix} r\dot{\psi} \\ 0 \\ 0 \end{pmatrix} \qquad (2.19)$$

In case of the steerable standard wheel, equation 2.19 can be used in the same way, if the fixed angle β is replaced by a function $\beta(t)$. This equation could also be applied to the spherical wheel (because of the forces which affect the wheel and change $\beta(t)$, only a linear velocity in the rolling direction exists). If the wheel is a castor wheel, the y-component of the velocity vector is depending on the angular velocity $\dot{\beta}$ and the length of the rod (see equation 2.20).

$$\begin{pmatrix} s(\alpha + \beta) & -c(\alpha + \beta) & -c\beta d \\ c(\alpha + \beta) & s(\alpha + \beta) & s\beta d \\ 0 & 0 & 0 \end{pmatrix} \begin{pmatrix} \dot{x} \\ \dot{y} \\ \dot{\theta} \end{pmatrix} = \begin{pmatrix} r\dot{\psi} \\ -d_c\dot{\beta} \\ 0 \end{pmatrix} \qquad (2.20)$$

The Swedish or Mecanum wheel is able to move in an omnidirectional way. Therefore, lateral movement of the wheel should be possible and can be calculated with equation 2.21.

$$\begin{pmatrix} s(\alpha + \beta + \gamma) & -c(\alpha + \beta + \gamma) & -c(\beta + \gamma)d \\ c(\alpha + \beta + \gamma) & s(\alpha + \beta + \gamma) & s(\beta + \gamma)d \\ 0 & 0 & 0 \end{pmatrix} \begin{pmatrix} \dot{x} \\ \dot{y} \\ \dot{\theta} \end{pmatrix}$$
$$= \begin{pmatrix} r\dot{\psi}\cos\gamma \\ r\dot{\psi}\sin\gamma + r_{pr}\dot{\psi}_{pr} \\ 0 \end{pmatrix} \qquad (2.21)$$

2.2.1 Kinematics of a differential drive vehicle

To calculate the kinematics of a differential drive, vehicle first the wheel types used have to be determined. This type of robot has two fixed standard wheels which are mounted on one axis. The kinematic center is located in the middle of the axis; the distance between the kinematic center and each wheel should be d. To solve the kinematics problem the coordinate system to define the parameters must be specified. The origin of this frame lies on the kinematic center. One solution for modelling the wheel configuration is to place the wheels on the y-axis of the coordinate frame. As shown in figure 2.5, the parameters are $\alpha_l = 90°, \beta_l = 0°, \alpha_r = -90°, \beta_r = 180°$. An example of a differential drive robot is MARVIN, the mobile vehicle of the University of Kaiserslautern (see figure 2.6).

Based on equation 2.19 one obtains for the left and the right wheel:

$$s(\alpha_l + \beta_l)\dot{x} - c(\alpha_l + \beta_l)\dot{y} - c\beta_l d\dot{\theta} = r_l\dot{\psi}_l$$
$$c(\alpha_l + \beta_l)\dot{x} + s(\alpha_l + \beta_l)\dot{y} + s\beta_l d\dot{\theta} = 0$$
$$s(\alpha_r + \beta_r)\dot{x} - c(\alpha_r + \beta_r)\dot{y} - c\beta_r d\dot{\theta} = r_r\dot{\psi}_r \qquad (2.22)$$
$$c(\alpha_r + \alpha_r)\dot{x} + s(\alpha_r + \beta_r)\dot{y} + s\beta_r d\dot{\theta} = 0$$

If the above mentioned parameters are inserted, the following equation will result:

$$\dot{x} - d\dot{\theta} = r_l\dot{\psi}_l$$
$$\dot{y} = 0$$
$$\dot{x} + d\dot{\theta} = r_r\dot{\psi}_r \qquad (2.23)$$
$$\dot{y} = 0$$

After solving the equation system on receives:

$$\begin{pmatrix} \dot{x} \\ \dot{y} \\ \dot{\theta} \end{pmatrix} = \begin{pmatrix} \frac{1}{2}(r_l\dot{\psi}_l + r_r\dot{\psi}_r) \\ 0 \\ \frac{1}{2d}(-r_l\dot{\psi}_l + r_r\dot{\psi}_r) \end{pmatrix} \qquad (2.24)$$

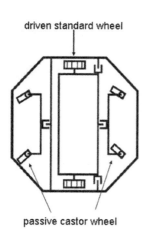

driven standard wheel

passive castor wheel

Figure 2.6 The autonomous vehicle MARVIN of the University of Kaiserslautern

2.2.2 Kinematics of an omnidirectional vehicle

To increase the mobility of a vehicle, an omnidirectional drive can be used. The climbing robot CROMSCI (see figure 2.7) of the University of Kaiserslautern, for example, is equipped with such a drive, in which 3 steerable standard wheels are mounted with an angle displacement of 120° between them (see figure 2.8). To set up the kinematics equations one can model the wheel configuration as shown in figure 2.5 with the kinematic center in the middle of the robot. The front wheel is located on the x-axis, the two rear wheels have a displacement to the front wheel of ±120°. As shown in figure 2.5 $\alpha_1 = 0°, \alpha_2 = 120°, \alpha_3 = -120°$. $\beta_{1,2,3}$ are the control parameters to determine the direction of the vehicle movements. The parameter d describes the distance between the wheel's contact point and the robot center.

For the navigation of CROMSCI it is necessary to calculate based on the desired linear and rotational velocities of the kinematic center $(\dot{x}, \dot{y}, \dot{\theta})^T$, the single wheel velocities and the orientations $(r_1\dot{\psi}_1, r_2\dot{\psi}_2, r_3\dot{\psi}_3, \beta_1, \beta_2, \beta_3)$.

Figure 2.7 The climbing robot CROMSCI (left) and the wheel settings (right)

Figure 2.8 Typical orientations of the 3 steerable wheels of an omnidirectional vehicle

Applying equation 2.19 for each wheel leads to the following equation systems:

$$s(\alpha_1 + \beta_1)\dot{x} - c(\alpha_1 + \beta_1)\dot{y} - d \cdot c(\beta_1)\dot{\theta} = r_1\dot{\psi}_1$$
$$s(\alpha_2 + \beta_2)\dot{x} - c(\alpha_2 + \beta_2)\dot{y} - d \cdot c(\beta_2)\dot{\theta} = r_2\dot{\psi}_2$$
$$s(\alpha_3 + \beta_3)\dot{x} - c(\alpha_3 + \beta_3)\dot{y} - d \cdot c(\beta_3)\dot{\theta} = r_3\dot{\psi}_3$$
$$c(\alpha_1 + \beta_1)\dot{x} + s(\alpha_1 + \beta_1)\dot{y} + d \cdot s(\beta_1)\dot{\theta} = 0$$
$$c(\alpha_2 + \beta_2)\dot{x} + s(\alpha_2 + \beta_2)\dot{y} + d \cdot s(\beta_2)\dot{\theta} = 0$$
$$c(\alpha_3 + \beta_3)\dot{x} + s(\alpha_3 + \beta_3)\dot{y} + d \cdot s(\beta_3)\dot{\theta} = 0$$

$$(2.25)$$

Based on the last 3 equations of 2.25, the steering angles $\beta_i, i = 1,2,3$ are determined:

$$c(\alpha_i + \beta_i) \cdot \dot{x} + s(\alpha_i + \beta_i) \cdot \dot{y} + d \cdot s(\beta_i) \cdot \dot{\theta} = 0$$
$$\Rightarrow c(\alpha_i) \cdot c(\beta_i) \cdot \dot{x} - s(\alpha_i) \cdot s(\beta_i) \cdot \dot{x}$$
$$+ s(\alpha_i) \cdot c(\beta_i) \cdot \dot{y} + c(\alpha_i) \cdot s(\beta_i) \cdot \dot{y} + d \cdot s(\beta_i) \cdot \dot{\theta} = 0$$
$$\Rightarrow c(\beta_i) \cdot (c(\alpha_i) \cdot \dot{x} + s(\alpha_i) \cdot \dot{y}) = s(\beta_i) \cdot (s(\alpha_i) \cdot \dot{x} - c(\alpha_i) \cdot \dot{y} - d \cdot \dot{\theta}) \quad (2.26)$$
$$\Rightarrow \tan(\beta_i) = \frac{s(\beta_i)}{c(\beta_i)} = \frac{c(\alpha_i) \cdot \dot{x} + s(\alpha_i) \cdot \dot{y}}{s(\alpha_i) \cdot \dot{x} - c(\alpha_i) \cdot \dot{y} - d \cdot \dot{\theta}}$$
$$\Rightarrow \beta_i = \text{atan2}\left((c(\alpha_i) \cdot \dot{x} + s(\alpha_i) \cdot \dot{y}), (s(\alpha_i) \cdot \dot{x} - c(\alpha_i) \cdot \dot{y} - d \cdot \dot{\theta})\right)$$

From equation 2.25, the angular velocity of the wheel $\dot{\psi}_i$ can be calculated using β_i:

$$\dot{\psi}_i = \frac{1}{r_i}\left(s(\alpha_i + \beta_i)\dot{x} - c(\alpha_i + \beta_i)\dot{y} - d \cdot c(\beta_i)\dot{\theta}\right) \quad (2.27)$$

2.2.3   Kinematics of a vehicle with Mecanum wheels

Another drive system suited for an omnidirectional vehicle are Mecanum wheels. Those are convex cylinders arranged in a 45° angle relative to the wheel plane. Two pairs of independently driven Mecanum wheels are sufficient to enable omnidirectional movement. The orientation of the rollers of the wheels lying on a common diagonal axis is equal. The rollers of the other two wheels are oriented in the opposite direction. In figure 2.9 (right side) 4 typical movements of the vehicle are shown. If all wheels move with the same velocity in the same direction, the robot drives straight ahead. The machine will turn if the right and left wheels move in opposite direction with the same velocity. A sideward motion is possible if the neighboring wheels move in opposite direction with the same velocity. A diagonal motion results if the two wheels on the diagonal move in the same direction with the same velocity. This type of drive was applied for the vehicles PRIAMOS of Prof. Dillmann's research group at the University of Karlsruhe (see figure 2.10).

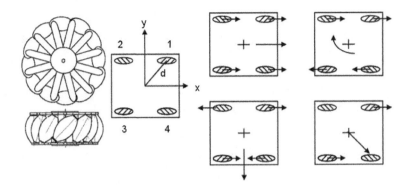

Figure 2.9 Schematic configuration of a Mecanum wheel

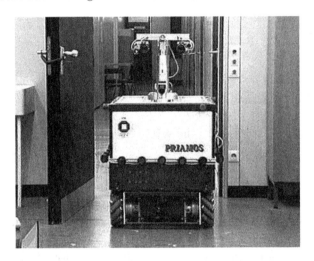

Figure 2.10 The mobile robot PRIAMOS of the University of Karlsruhe driven by Mecanum wheels [DKWW95] (courtesy of Prof. Dillmann, TH Karlsruhe)

To set up the kinematic equation, the parameters (α,β,γ) for each wheel must be determined. The order of the wheels is shown in figure 2.9. The parameters for four wheels are:

$$\alpha_1 = 45°, \qquad \beta_1 = 45°, \qquad \gamma_1 = -45°$$
$$\alpha_2 = 135°, \qquad \beta_2 = -45°, \qquad \gamma_2 = 45°$$
$$\alpha_3 = -135°, \qquad \beta_3 = 225°, \qquad \gamma_3 = -45°$$
$$\alpha_4 = -45°, \qquad \beta_4 = 135°, \qquad \gamma_4 = 45°$$

Using equation 2.21, and supposing all driven wheels have the same radius r, the same distance d from the kinematic center and the above mentioned parameters for α,β,γ are inserted, we receive:

$$s(45°)\dot{x} - c(45°)\dot{y} - d\dot{\theta} = r \cdot c(-45°)\dot{\psi}_1 \tag{2.28}$$

$$s(135°)\dot{x} - c(135°)\dot{y} - d\dot{\theta} = r \cdot c(45°)\dot{\psi}_2 \tag{2.29}$$

$$s(45°)\dot{x} - c(45°)\dot{y} - d\dot{\theta} = r \cdot c(-45°)\dot{\psi}_3 \tag{2.30}$$

$$s(135°)\dot{x} - c(135°)\dot{y} - d\dot{\theta} = r \cdot c(45°)\dot{\psi}_4 \tag{2.31}$$

Based on this equation system, the velocities $\dot{x},\dot{y},\dot{\theta}$ of the kinematic center can be calculated:

$$\dot{x} = \frac{r}{4}(\dot{\psi}_1 + \dot{\psi}_2 + \dot{\psi}_3 + \dot{\psi}_4)$$

$$\dot{y} = \frac{r}{4}(-\dot{\psi}_1 + \dot{\psi}_2 - \dot{\psi}_3 + \dot{\psi}_4) \tag{2.32}$$

$$\dot{\theta} = \frac{r}{d\sqrt{2}}(\dot{\psi}_1 - \dot{\psi}_2 - \dot{\psi}_3 + \dot{\psi}_4)$$

The velocity vector of the kinematic center can be determined as:

$$\begin{pmatrix} \dot{x} \\ \dot{y} \\ \dot{\theta} \end{pmatrix} = \frac{r_{\text{wheel}}}{4} \cdot \begin{pmatrix} 1 & 1 & 1 & 1 \\ -1 & 1 & -1 & 1 \\ C & -C & -C & C \end{pmatrix} \cdot \begin{pmatrix} \dot{\psi}_1 \\ \dot{\psi}_2 \\ \dot{\psi}_3 \\ \dot{\psi}_4 \end{pmatrix} \tag{2.33}$$

with $C = \frac{2\sqrt{2}}{d}$. In order to assume the absence of slip the following must hold: $\dot{\psi}_4 = \dot{\psi}_1 + \dot{\psi}_2 - \dot{\psi}_3$

2.2.4 Pose calculation based on velocities

Using equation 2.34 and introducing the time interval Δt, the incremental paths can be determined for 2D navigation as:

$$\begin{pmatrix} \Delta x \\ \Delta y \\ \Delta \theta \end{pmatrix} = \begin{pmatrix} v_x \\ v_y \\ \omega \end{pmatrix} \cdot \Delta t \tag{2.34}$$

Assuming the velocity $v = (v_x(t),v_y(t))_A$ given in a robot fixed coordinate frame (x_A,y_A), angular velocity $\omega(t)$ and the robot pose (x,y,θ) in the world coordinate system given, one can compute the trajectory as:

$$\theta(t) = \int_0^t \omega(\tau)d\tau + \theta_0 \tag{2.35}$$

$$x(t) = \int_0^t \dot{x}(\tau)d\tau + x_0 \tag{2.36}$$

$$y(t) = \int_0^t \dot{y}(\tau)d\tau + y_0 \tag{2.37}$$

Vehicle velocities are $\dot{x}$ due to the x-axis, $\dot{y}$ due to the y-axis and $\omega = \dot{\theta}$ around z-axis.

2.3 Geometrical solution for vehicle kinematics

As presented above, the drive kinematics can be calculated by using the wheel parameters. Nevertheless, geometrical solutions are commonly used if the robot is moving only in 2D space. That is because it is much easier to calculate and comprehend. In the following, some standard vehicle concepts well known from literature will be presented and the kinematics problem will be solved geometrically.

2.3.1 Differential drive

A differential drive setup consists of two independently driven wheels and thus only circular arc trajectories are possible. Therefore, two special cases occur: $R = 0$ and $R = \infty$. The former results in a rotation on the spot while the latter results in a straight route. Hence differential drives possess two independent DOF. Midway between the driven wheels the kinematic center is situated. An exemplary robot system based on differential drive is the service robot ARTOS developed at the RRLab at the University of Kaiserslautern (see figure 2.11).

Given the single wheel velocities v_l and v_r of the left respectively right wheel and a time step Δt, the length of the driven ways for each wheel and the kinematic center Δs_m can be calculated (see figure 2.12):

$$\Delta s_l = v_l \cdot \Delta t$$
$$\Delta s_r = v_r \cdot \Delta t$$
$$\Delta s_m = (\Delta s_l + \Delta s_r)/2 \tag{2.38}$$

Figure 2.11 Differential drive robot ARTOS of the University of Kaiserslautern

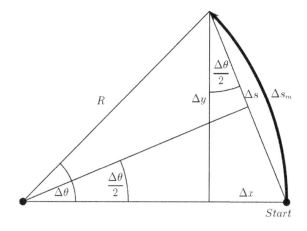

Figure 2.12 Geometrical solution of the differential drive kinematic

Based on this and the distance d between a wheel and the kinematic center, the radius R can be derived:

$$R = d \cdot \frac{\Delta s_r + \Delta s_l}{\Delta s_r - \Delta s_l} \qquad (2.39)$$

Keep in mind that we drive a left curve. Otherwise the radius will be negative (curve to the right). The change in orientation can be calculated using

$$\Delta\theta = \frac{\Delta s_m}{r_m} \qquad (2.40)$$

$$= \frac{\Delta s_r - \Delta s_l}{2 \cdot d} \qquad (2.41)$$

For calculating translation changes one need the length of

$$\Delta s = 2 \cdot r_m \cdot \sin\left(\frac{\Delta\theta}{2}\right) \tag{2.42}$$

Based on this and the robot's orientation θ_0 at the starting position, the final changes can be derived:

$$\Delta x = \Delta s \cdot \cos\left(\frac{\Delta\theta}{2} + \theta_0\right)$$
$$\Delta y = \Delta s \cdot \sin\left(\frac{\Delta\theta}{2} + \theta_0\right) \tag{2.43}$$

2.3.2 Tricycle drive

This very common setup is based on a three wheel concept, see figure 2.13. The steerable front wheel is driven while the two wheels in the back are free-wheeling. Again this simply results in circular arc trajectories, but this time the minimum radius is bigger than zero. Therefore, the robot is unable to turn on the spot. The kinematics can be derived in an analogue way to the one of Ackermann steering that will be presented in the next paragraph.

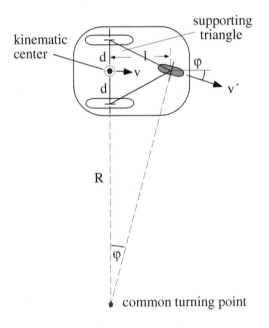

Figure 2.13 Tricycle kinematics

Suppose the steering angle φ and the linear velocity v' of the driven wheel are given, one can calculate the Radius R and the velocity v of the kinematic center with wheel distance d and axis distance l:

$$R = l \cdot \cot \varphi$$
$$v' = \frac{v}{\cos \varphi} \tag{2.44}$$

These two values can be used to derive the velocities of both rear wheels:

$$v_l = v \cdot \frac{R+d}{R}$$
$$v_r = v \cdot \frac{R-d}{R}$$

Based on these values the vehicles kinematics as shown before in section 2.3.1 can be calculated.

2.3.3 Ackermann steering

This specific type of drive system is mostly found in the field of automotive applications. It consists of a fixed axle and another one connecting the parallel steered wheels. In case the driven wheels are connected to the fixed axle, a differential has to be included in the setup in order to allow for curved trajectories. If the steered wheels are driven the differential is obsolete. Ackermann drive setups possess three degrees of freedom, however they are not independent. The control of an Ackermann steering is complex, as all car owners might already have experienced themselves. Nevertheless, those setups find various application in the field of robotics, for instance in the commercial outdoor platform RobuCar. The desired drive speed is denoted v_D while v_{RR} and v_{LR} denote the rear left and right wheel speed. v_{RF}, v_{LF} denote the respective speeds for the front axles while l denotes the length of the vehicle and d the distance between wheel and kinematic center.

Based on the introduces parameters and steering angle φ one can derive the circle's radius

$$R = \frac{l}{\tan \varphi} \tag{2.45}$$

and finally the four wheel velocities

$$v_{LR} = \frac{(R-d) \cdot v_D}{R}$$

$$v_{RR} = \frac{(R+d) \cdot v_D}{R}$$

$$v_{LF} = \frac{\sqrt{(R-d)^2 + l^2} \cdot |\tan\varphi|}{l} \cdot v_D \qquad (2.46)$$

$$v_{RF} = \frac{\sqrt{(R+d)^2 + l^2} \cdot |\tan\varphi|}{l} \cdot v_D$$

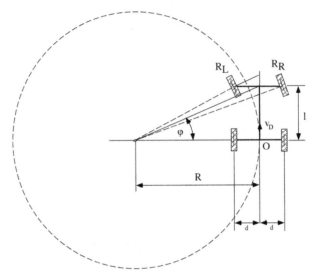

Figure 2.14 Kinematics of Ackermann steering

2.3.4 Double Ackermann steering

In a double Ackermann steering both axles are steerable, see figure 2.15. It is obvious that such a setup is kinematically even more complex and problematic than the Ackermann setup already discussed. When a curve is steered, two rotation points of the robot motion will occur. This yields slip of the single wheels. This way positioning the robot becomes quite complicated. However, the advantages are a smaller turning circle as well as the possibility of sideward motions in case both axles are steered in parallel. Especially in off-road applications (e.g. robot RAVON in figure 2.16), the errors of this configuration are lower than those of the interaction between vehicle and terrain.

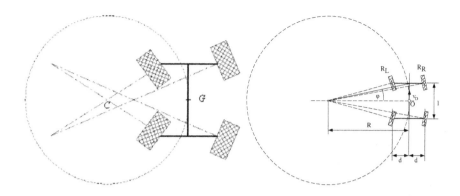

Figure 2.15 Double Ackermann steering with slip (left) and with common pivot point (right)

Figure 2.16 Robot RAVON of University of Kaiserslautern at an early stage

Using known length l, distance d, desired velocity v_D and steering angle φ as already presented in section 2.3.3, one can calculated the radius R and wheel velocities:

$$R = \frac{l}{\tan \varphi}$$

$$v_{LR} = \frac{\sqrt{\left(\frac{R}{2} - d\right)^2 + \frac{l}{4}^2} \cdot 2|\tan \varphi|}{l} \cdot v_D$$

$$v_{RR} = \frac{\sqrt{\left(\frac{R}{2} + d\right)^2 + \frac{l}{4}^2} \cdot 2|\tan \varphi|}{l} \cdot v_D \qquad (2.47)$$

$$v_{LF} = \frac{\sqrt{\left(\frac{R}{2} - d\right)^2 + \frac{l}{4}^2} \cdot 2|\tan \varphi|}{l} \cdot v_D$$

$$v_{RF} = \frac{\sqrt{\left(\frac{R}{2} + d\right)^2 + \frac{l}{4}^2} \cdot 2|\tan \varphi|}{l} \cdot v_D$$

2.3.5 Synchro drive

The main feature of this type of drive system is that all wheels are equally steered, see figure 2.17. The minimum amount of motors required is two: the first one drives the wheels by either a chain or a belt, the second one is responsible for controlling the steering angle. Thus, all wheels are always rotating equally fast and are facing the same way. A vehicle equipped with such a drive setup is able to reach any given point in a plane but is limited to two DOF since it cannot rotate. This is of special importance for the layout of the sensors since they will not necessarily face the direction of the movement. An example for a robot that relies on such a drive setup is the industrial service robot Viper [GRD98].

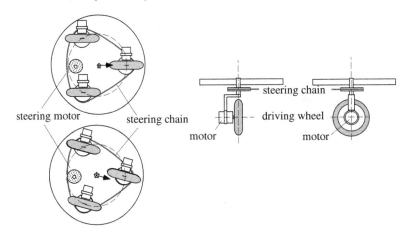

Figure 2.17   Schematics of a basic synchro drive

The deduction of the drive kinematics is straightforward because only basic trigonometry is applied to derive the solution. Let Δs denote the length of the travelled path of the driven wheels while φ is the steering angle. Thus we receive

$$\Delta x = \Delta s \cdot \cos \varphi$$
$$\Delta y = \Delta s \cdot \sin \varphi \tag{2.48}$$

If used the other way around, the above equation will determine the desired parameters:

$$\Delta s = \sqrt{\Delta x^2 + \Delta y^2}$$
$$\varphi = \arctan \frac{\Delta y}{\Delta x} \tag{2.49}$$

As already mentioned the synchro drive is unable to perform rotations. This, however, implies that $\Delta \varphi = 0 = const$ holds true!

2.3.6   Omnidrive

An omnidrive system consists of a minimum of two independently steered
wheels with one or more free-wheeling passive wheels serving as supporting
wheels. Thus the vehicle is able to move in a plane with three DOF.

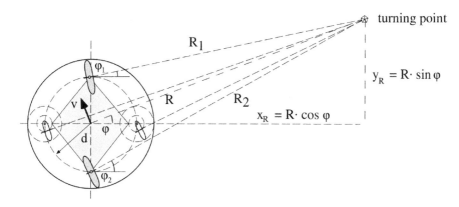

Figure 2.18   Omnidrive kinematics

Motion radius R, steering angle φ, linear velocity v of the kinematic center and wheel distance d are given. To calculate the single wheel orientations
and velocities that are needed, the following helping distances can be used:

$$
\begin{aligned}
x_R &= R \cdot \cos \varphi \\
y_R &= R \cdot \sin \varphi
\end{aligned}
\tag{2.50}
$$

Based on this one can calculate the wheel parameters of the first

$$
\begin{aligned}
\varphi_1 &= \arctan \frac{y_R - d}{x_R} \\
R_1 &= R \cdot \frac{\cos \varphi}{\cos \varphi_1} \\
v_1 &= v \cdot \frac{R_1}{R}
\end{aligned}
\tag{2.51}
$$

and of the second wheel:

$$
\begin{aligned}
\varphi_2 &= \arctan \frac{y_R + d}{x_R} \\
R_2 &= R \cdot \frac{\cos \varphi}{\cos \varphi_2} \\
v_2 &= v \cdot \frac{R_2}{R}
\end{aligned}
\tag{2.52}
$$

Regarding a generalized omnidrive one needs different parameters which describe the kinematic setup. At the distance d from the kinematic center of a vehicle there is a wheel at coordinates (x_i, y_i) with radius r. At an orientation φ with respect to the x-axis the vehicle drives at a velocity v and has an angular speed of ω as shown in figure 2.19.

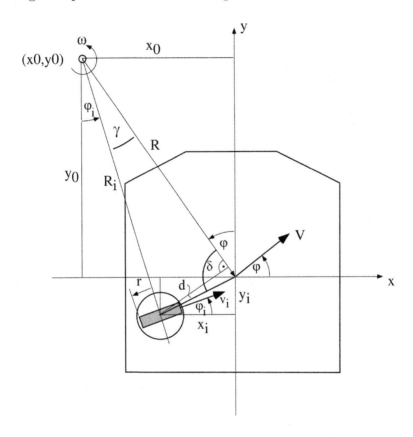

Figure 2.19 Generalized omnidrive

Given the linear velocity v, angular velocity ω, angle φ and the coordinates (x_i, y_i) of a driving wheel number i one can calculate the pivot point with coordinates (x_0, y_0) and the motion radius R:

$$R = \frac{v}{\omega}$$
$$x_0 = -R \sin \varphi \qquad\qquad (2.53)$$
$$y_0 = R \cos \varphi$$

By using these values the steering angle φ_i and the angular velocity ψ_i of wheel i can be derived:

$$R_i = \sqrt{(x_0 - x_i)^2 + (y_0 - y_i)^2}$$
$$\psi_i = \omega \frac{R_i}{r} \tag{2.54}$$
$$\varphi_i = \sin^{-1}\left(\frac{x_0 - x_i}{R_i}\right)$$

Driving straight on means $R \to \infty \implies \varphi_i = \varphi \; ; \; \psi_i = v/r$.

2.4 Applying mobile robot kinematics

With the kinematic models obtained above, it is now possible to solve simple navigation tasks. These models are the basis for any mobile robot application. The first problem to be solved is localization. Based on the angular velocities of the wheels and the kinematic parameters of the vehicle, one can determine the robot pose on a 2D plane. The resulting pose is only a first estimation because of slip effects of the wheels.

In figure 2.20, a typical scenario is presented in which a differential drive robot is supposed to move along a square of $10\,\text{m} \times 10\,\text{m}$. The wheel velocities are calculated due to its kinematics and the desired path. These velocities could directly be used for the closed-loop controllers, which determine e. g. the power of the motors. The ellipses show the area in which the real position of the robot could be for each step of movement. This error is dependent on the kinematics of the vehicle. One can observe that the size of the area is increasing with movement, because of the summation of slippage error. It is also shown that the distance error is smaller than the rotational error for a differential drive robot. This results in the elliptic shape of the areas. The orientation of the different ellipses depends on the orientation of the robotic system. This so-called dead reckoning localization is sufficient for short paths but not precise enough for navigation. In chapter 4 different methods for solving the localization problem will be introduced.

The kinematics could also be used to determine the DOF (degree of freedom) of the robot. In [SN04] the DOF are separated in degree of mobility and degree of steerability. The mobility describes possible independent motions based on changes to wheel velocities which are not restricted by kinematic constraints like the sliding condition. In the case of a differential drive robot the mobility degree is 2, because a movement along the robots' y-axis is restricted. Not all three motions of a vehicle in a plane (x-direction, y-direction, rotation) are possible at the same time. The steerability indicates the number of steerable wheels, which could be independently controlled. In case of a differential drive robot the steerability is 0, in case of a tricycle it is 1. The summation of both degrees leads to the maneuverability of the robot system.

Overall, kinematics is the foundation for solving different problems like localization, navigation, or SLAM, which will be presented in the next chapters.

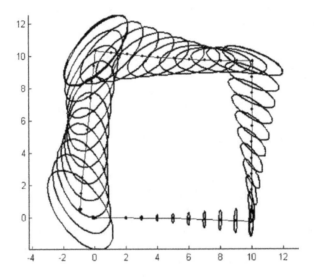

Figure 2.20 Typical error caused by odometry using a differential drive

3 Sensors

This chapter introduces sensors often usedcome on autonomous mobile vehicles. In general, a sensor or a sensor system transforms different kinds of physical values (e. g. a force or a velocity) into electrical signals. One can distinguish sensors according to the integration level (see also figure 3.1):

Basic sensor: measurement and transformation of the physical signals,

Integrated sensor: basic sensor with signal processing including amplification, filtering, linearization and normalization,

Intelligent sensor: integrated sensor with computer-controlled analysis of the processed signal.

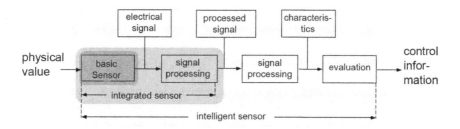

Figure 3.1 Sensor integration levels

Another way of classifying sensors often used in literature is the separation into proprioceptive sensors and exteroceptive sensors. Proprioceptive sensors measure internal states of a robot (e. g. wheel velocity or acceleration), while exteroceptive sensors observe the state of the robot in relation to its environment (e. g. distance to an obstacle or object identification).

On the other hand, one can also distinguish between active and passive sensors. Active sensors stimulate the environment and analyze the responding signal (e. g. ultrasound sensors, laser scanners) while passive sensors measure a present signal (camera, microphone).

All sensors share an inability to measure their respective variable perfectly. There always is a measurement error, depending on the measurement principle.

In the following, sensors that are useful for autonomous land vehicles are summarized according to perception characteristics and operating principles. It makes no sense to describe specific products on the market in this section, since they change continuously.

3.1   Tactile sensors

Tactile sensors detect physical contact between the vehicle and an obstacle. This type of sensor is often used in simple mobile robots as a cheap possibility to describe the environment of the robot. On the other hand, tactile sensors are often used to fulfill safety requirements imposed on a robot: physical contact of the vehicle with an obstacle causes an emergency stop.

3.1.1   Switches

Operated by a force larger than a defined minimal force, a spring loaded contact, which closes or opens an electric circuit, induces a signal. This might be used to detect more than an unwanted contact with the environment.

3.1.2   Bumper

As a last resort before crashing into an obstacle a bumper activates an emergency break and consumes some part of the impulse like passive bumpers on cars. This is done by a more or less elastic shield, fastened by springs to the vehicle chassis. Switches detect the deformation of the shield (see figure 3.2).

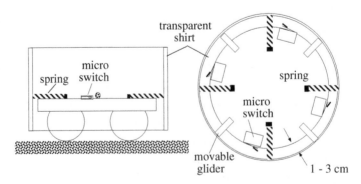

Figure 3.2 Sketch of a simple bumper fixed to the robot chassis (sideview and bird's eye view)

3.1.3   Force sensors

Typically, force sensors use the physical deformation of elastic materials. This deformation changes properties of the material like electrical resistance or capacity. The parameters of a force sensor are the minimal detectable force, the maximal allowed force and the speed at which a changing force can be measured. In wheel-driven vehicles, force sensors fixed on the wheels are often used to prevent tilting over. Also slipping of the vehicle can be detected with the help of force sensors.

For example, the climbing robot CROMSCI of the University of Kaiserslautern possesses 3 steerable wheels. In each wheel a 3-component force sensor is integrated which measures contact forces in x-, y-, and z-direction, see figure 3.3 and 3.4. The maximum forces in x- and y-direction are 150 N while in z-direction 1500 N could be detected with a resolution of 1 N.

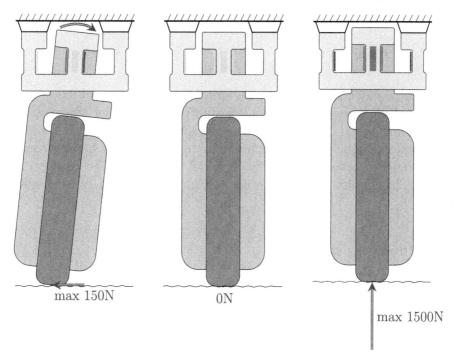

Figure 3.3 Measurement principle of wheel-mounted force sensors. Strain gages are glued to specific parts of the sensor box. The forces in the wheels lead to a deformation of these parts which cause a change in the resistance of the strain gages. The measured fall of voltage is proportional to the change of forces.

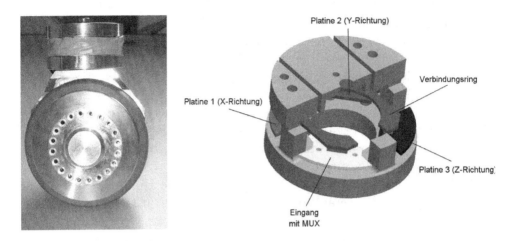

Figure 3.4 The driven wheel of CROMSCI equipped with a force sensor (left) and the force sensor with integrated electronics (right)

3.2 Pose measurement

The pose of a vehicle is its position and orientation with respect to a world coordinate system. There are different possibilities to determine this pose. Most techniques are based on odometry sensors, landmarks, magnetic compasses or inclinometers. The sensors which belong to these classes have in common that they are not precise enough to solve the pose estimation problem independently. In chapter 4, some methods of improving postion estimates are shown.

3.2.1 Odometry sensors

A robot's trajectory derived from the summation of wheel speeds is called odometry. However without orientation sensor, this odometry works only in 2D.

Wheel encoders are mainly used for the determination of the wheel speed. Based on the measurement principle one can discern magnetic and optical wheel encoders. Magnetic wheel encoders measure the number of magnetic pole changes of magnets at the perimeter of a disk either by hall effect sensors or by the voltage produced by an electric generator turned by the wheel (tacho generator).

Optical wheel encoders measure distances by the number of ticks produced by a grid passing a light barrier. There are two grids: one fixed to the vehicle chassis, the other one turning with the wheel. Hence, the movement of the vehicle might be tracked simply counting the number of ticks at the wheels resulting in the distance travelled.

Let n be the number of ticks measured and n_0 the number of ticks for a full revolution of the wheel with radius r. The distance travelled then is $s = 2\pi r \cdot \frac{n}{n_0}$.

As presented in figure 3.5, the encoder is also capable of measuring the direction of the rolling wheel. Two grids, fixed to the vehicle chassis $\left(\left(\frac{n+1}{4}\right)\right)$ grid constants apart, emit two signals with a phase difference of $90°$ from which the direction may be derived. Typical wheel encoders have grids of 4096 equidistant transparent and nontransparent areas. These encoders have a resolution of 12 Bit or $0.1°$. The calculations towards the vehicle pose are to be found in chapter 4.

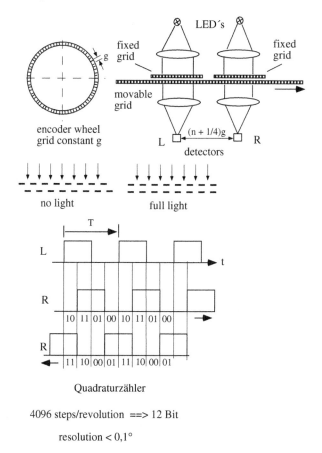

Figure 3.5 Measurement principle of a wheel-encoder

3.2.2   Compass

A magnetic compass can be used to keep a direction, measuring the horizontal component of the earth's magnetic field assumed to be constant in its direction over short distances. Apart from the classical compass as used on ships for hundreds of years, this can be done using a ferrite slab with a rectangular hysteresis loop. Any external magnetic field in the long direction of the slab shifts the magnetization where the slab goes into saturation.

In a coil wound around the slab driven by a varying current the point of saturation shows up in a voltage peak induced in the coil by the sudden change of magnetization. Driving the coil with a time varying current ramp, the hysteresis loop is scanned periodically. The component of the field in the direction of the slab can thus be measured as shown in figure 3.6. Two crossed slabs give the direction of the earth's magnetic field as seen from the vehicle. In indoor applications, the magnetic field in the room is mostly disturbed by metallic parts in concrete walls and ceilings.

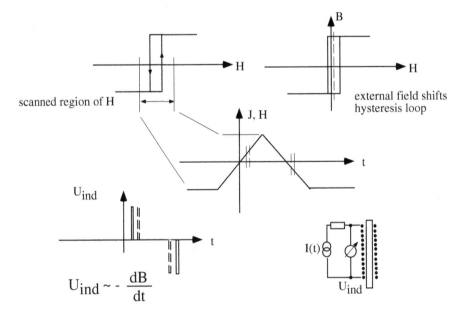

Figure 3.6   Principle of the flux-gate sensor

3.2.3   Inclinometers

In order to measure whether an autonomous vehicle moves up or down a steep incline or shifts to a side, with the danger of toppling or falling to the side, the inclination must be measurable in two axes. This can be done using a droplet of liquid under the influence of gravity as shown in figure 3.7. In a transparent hemispherical plastic bowl filled with oil, a droplet of water forms a lens at the bottom, because water is more dense than oil. The light of a LED is projected by that lens onto a CCD-matrix. Any inclination shifts the droplet and so the image of the LED on the CCD-matrix, indicating the amount of inclination. An oil of suitable viscosity damps the movement of the water droplet to damp jitter from movements over uneven ground.

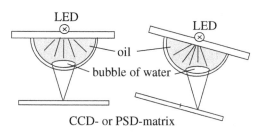

Figure 3.7 Measuring principle of an optical inclinometer

Figure 3.8 shows the principle of another device; a dielectric liquid drop floats between the plates of a capacitance. Any tilt shifts the drop to the side and changes the capacitances in a quad capacitance measuring device. Typical parameters for a specific sensor of this type are also shown.[1]

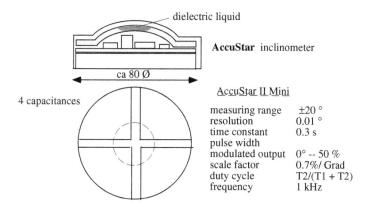

Figure 3.8 Measurement principle of a capacitive inclinometer

---

[1] Althen GmbH, Frankfurterstr. 150–152, D65779 Kelkheim, Germany; http://www.althen.de/neigungssensoren

3.3 Sensors for inertial systems

Determining the pose of a vehicle may be done by measuring only accelerations and turning rates and integrating these signals. These measurements
are independent from any disturbances of the outside world.

3.3.1 Acceleration sensors

Accelerometers measure the component of the acceleration in one direction.

Apart from a crash the accelerations in a vehicle driving around are in
general a lot smaller than the earth acceleration of $1\,g$.

Driving through a curve in a car will result in accelerations in the magnitude of approximately $0.1\,g$. Measuring devices found in an airbag system
can also be used for autonomous vehicles.

An example is the AD-XL 105 by Analog Devices.[2] It is a device fab-
ricated as a MEMS, a micro electro-mechanical system, etched out of silicon
and housed in a 14 pin DIP (dual inline package) like a small integrated
circuit.

Its parameters are

AD XL 105

measuring range	$\pm\,5\,g$
smallest detectable acceleration	$0.02\,g$
operating voltage	$5\,V$
output	$0.25\,V/g$ (analog)
bandwidth	$10\,kHz$

The device includes an uncommitted amplifier to amplify the output A
by the quotient of two external resistors: output $= (R_1/R_2) \cdot A \cdot 0.25\,V/g$
with acceleration A.

There are other sensors available that can measure more than one axis
at the same time. An example for such a sensor is presented in figure 3.9.
The AD-XL 202 is a two axis system by Analog Devices housed in a 14 pin
DIP with parameters

AD XL 202

measuring range	$\pm 5\,g$
smallest detectable acceleration	$0.01\,g$

---

[2] http://www.analog.com/en/mems-and-sensors/imems-accelerometers/

operating voltage	5 V
output	2 PWM TTL signals
pulse period	1–10 ms
measured acceleration A	$A = \frac{(T_2/T_1)-0.5}{12.5}$

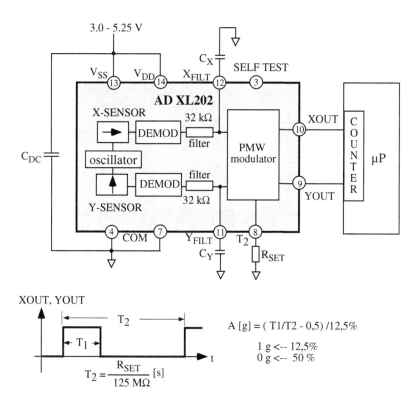

Figure 3.9 Two axis accelerometer AD XL 202

The pulse width is externally adjustable with 1 g equalling 12,5% and 0 g equalling 50%. Two built in filters at the output of the demodulator integrate the measured different signals. The integration time constant is set with the capacitors C_x and C_y. The period of the pulse width modulator at the output is set by resistor R_{SET}. Figure 3.10 shows a MEMS-version of this accelerometer.

Nowadays, usually micro electro-mechanical systems (MEMS) are used. They integrate complex electro-mechanical systems into bulk silicon. Silicon is a material with good properties for this type of application and it is very well understood in its physical and crystallographic properties from decades of producing integrated circuits. The same technique that is used to form

ICs may also be applied to etch out free swinging beams of silicon anchored
to the bulk material at a few points only.

µ machined two axis accelerometer

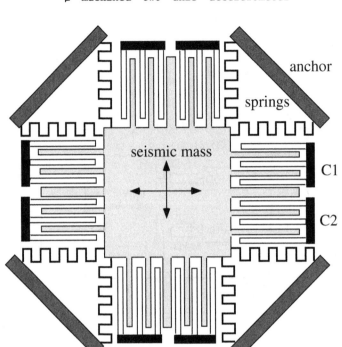

Example: Analog Devices AD XL 202

measuring range	± 2 g
smallest detectable acceleration	1 mg
operating voltage	5 V
housing	14 DIP
output	2 PWM TTL signals
pulse period	1-10 ms externally adjustable

Figure 3.10   Two axis accelerometer

The measurement principle here is a pair of capacitors to measure very
small distance variations under acceleration. A central difficulty with acceleration sensors is given by earth acceleration which in most cases is far
larger than the acceleration of the vehicle to be measured. Therefore, the
orientation information is crucial to remove unwanted gravity forces from
the measurement results.

Figure 3.11 depicts the principle. A mass anchored to the bulk by plate springs is forced into oscillations by a comb of slabs acting as capacitors. Typical dimensions of the slabs are length $s = 125\,\mu m$, thickness $b = 1\,\mu m$ and distance to the next slab $d = 1\,\mu m$. The force is multiplied by the number of combs gripping into each other. It acts in one direction only, but activating two combs at each side of the masses by alternatively switching the voltage driving the combs will set the system into oscillations. Typical frequencies are in the order of approx. $20\,\text{kHz}$. The forces are small, but as the damping of the system is very small, they are sufficient to drive the block into oscillations with voltages of $5\,\text{V}$.

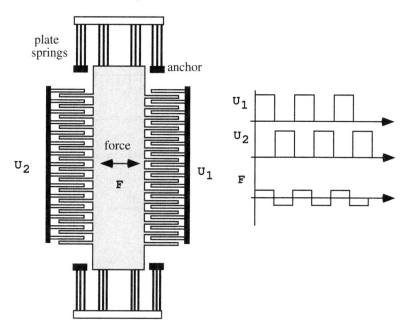

Figure 3.11 Forced oscillations

To measure small deviations down to $1\,\text{nm}$ the slabs may be used as differential capacitors as shown in figure 3.12.

Driving the capacitors by two voltages with $180°$ phase shift, the resulting signal after synchronous demodulation is proportional to the difference in capacitances and measures the deviation Δd. Since the measuring frequency is much larger than the frequency driving the oscillation they do not interfere with each other.

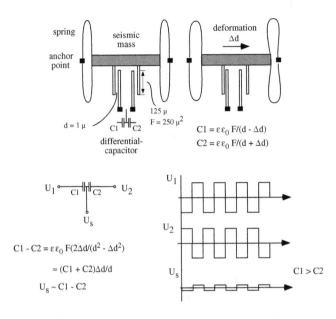

differential capacitor

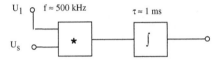

measuring of U_s by synchroneous demodulation with U_1 and integration

Figure 3.12 Differential capacitor

3.3.2 Turning rate sensors

Angular velocity sensors are intended to measure the turning rate of the vehicle. Their precision need not be too high: it is neither necessary to see the rotation of the earth with 15°/h at a pole, nor the slow rotation of the short arm of a clock of 30°/h or 0,008°/s if the position and orientation has to be kept for minutes only until the vehicle can reorient itself from landmarks again.

Cheap sensors for this application are again MEMS devices. Figure 3.13 shows such a sensor. Two seismic masses are forced into oscillations by comb drivers, operating 180° out of phase. Inside each mass a measuring comb forming the differential capacitors C_a, C_b, C_c and C_d respectively measure any deviation due to the Coriolis force and the amplitude of the driving oscillation.

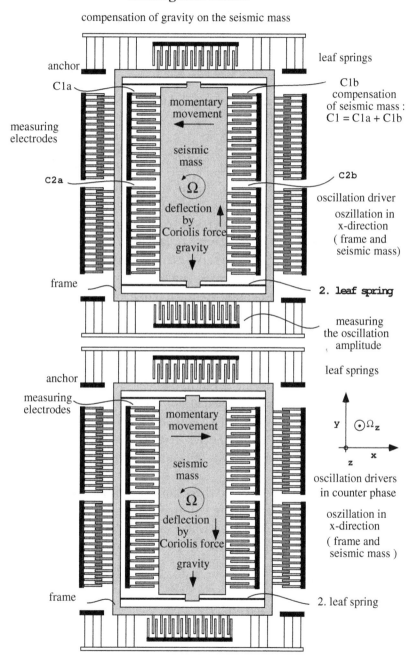

turning rate sensor

compensation of gravity on the seismic mass

Figure 3.13 Turning rate sensor

The amplitude of the driving oscillation is $A\omega \sin \omega t$ using the differential capacitors as $(C_a + C_c) - (C_b + C_d)$, while the amplitude of the turning rate Ω is measured through $\Omega A\omega \sin \omega t$ using $(C_a + C_b) - (C_c + C_d)$ in a synchronous detector.

An example of a turning rate sensor is ADXRS150 from Analog-Devices.[3]

ADXRS150

measuring range	$\pm 150°/s$ i.e. $\pm 2,62\,\text{rad/s}$
non linearity	0,1%
acceleration influence	$0.023°/s/g$
temperature influence	15% or $21.7°/s(-40°$ C to $+85°\text{C})$, $0.17°/s/°\text{C}$

The latest MEMS devices include 3-axis accelerometers with dimensions of $4\,\text{mm} \times 4\,\text{mm} \times 1.45\,\text{mm}$ in a 16-pin plastic lead frame chip scale package (LFCSP) as for instance the ADXL330 from Analog Devices. It offers a measuring range of $\pm 3\,\text{g}$ and a sensitivity of $300\,\text{mV/g}$. Packing a three-axis accelerometer and three turning rate sensors into one package gives a six-axis movement sensor, the ADIS16355 from Analog Devices.

In order to find the position and orientation of a vehicle without reference to external landmarks there must be precise sensors on board to measure the turning of the vehicle using inherent physical properties. If the position and orientation have to be maintained for longer times, the turning rate sensors must be precise – and consequently be expensive.

Laser Gyros use the so-called Sagnac effect. It exploits the fact that the speed of light stays constant regardless of the velocity of the sender. In a material with the refractive index $n(\lambda)$ and the speed of light in vacuum c_0, light travels at a velocity of $c = c_0/n(\lambda)$ irrespective of the velocity of the material itself. A comparatively cheap measuring device based on this effect is a laser fiber coil gyro.

Two light beams run in opposite directions in a fiber coil: one clockwise the other one counterclockwise. If the coil itself turns, a phase shift between entrance and exit of $\varphi = 2\omega LD/(\lambda c)$ will occur with L being the fiber length, D the diameter and ω the turning rate of the coil. Figure 3.14 illustrates the principle.

The system's specifications are:

Hitachi Optical Fiber Gyroscope

maximum input rotation rate	$\pm 100°/s$
minimum detectable rotation rate	$\pm 0,01°/s$
zero point drift	$\leq 10°/h$

---

[3] http://www.analog.com/en/mems-and-sensors/imems-gyroscopes/

dimensions	$100\,\text{mm}\times100\,\text{mm}\times60\,\text{mm}$
weight	$0.5\,\text{kg}$
analog output	$\pm2.5V$ at $350\,\text{mA}$
digital output	TTL, 9600 Bit/s
answer time	$10\,\text{ms}$
warm up time	$\leq1\,\text{min}$

A zero point drift of $10°/h$ is just the rotation of the earth at $40°$ latitude. A rotation rate of $0.01°/s$ equals $36°/h$ and is enough to detect the earth rotation. In order to derive the turning angle the signal from the rate sensor has to be integrated:

$$\psi(t) = \int_0^t \omega(\tau)d\tau \tag{3.1}$$

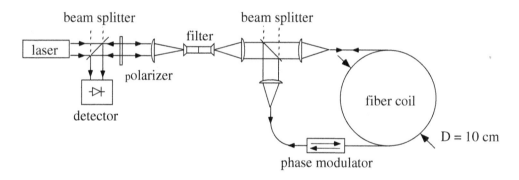

Figure 3.14 Principle of Hitachi Fiber Coil Gyro

Mechanical Gyros make use of the fact that a system with a constant angular momentum isolated from external torques keeps its angular momentum and thus its direction.

A torque $\vec{D}$ acting upon a system with angular momentum $\vec{J}$ induces a precession ω_p with $\vec{J}\omega_p = \vec{D}$ in the system.

Its direction is perpendicular to the plane spanned by $\vec{D}$ and $\vec{J}$.

Its magnitude is the product of the magnitudes of $\vec{D}$ and $\vec{J}$ and the sinus of the included angle γ as shown in figure 3.15 – the cross-product of the vectors of torque and angular momentum.

This is directly exploited in mechanical gyros – costly but precise. For autonomous vehicles, maintaining a desired direction is necessary for rather short periods only when there is no external reference available. In this case turning rate sensors based on the Coriolis force may be used: A swinging mass represents an angular momentum too and a forced precession ω_p is translated into a force F perpendicular to the swinging motion and the forced precession, as shown in the same figure as above and exploited in MEMS turning rate sensors described above.

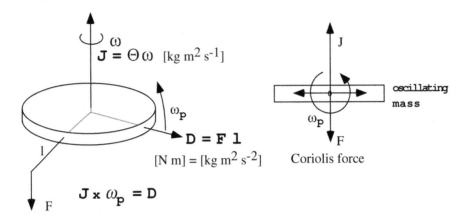

Figure 3.15 Precession and Coriolis force

Murata Gyro Star[4] An explicit Coriolis force based sensor is built around a triangular quartz prism set into vibrations like a quartz slab in an electronic clock. Any turning along the main axis causes the prism to swing in other directions as well.

Two electrodes measure these vibrations. Their signal is proportional to the turning rate as shown in figure 3.16.

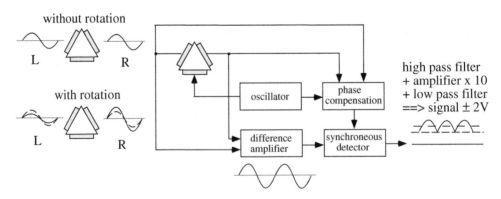

Figure 3.16 Principle of the Murata gyro compass

---

[4] http://www.murata.com/catalog/s42e.pdf

Typical specifications are listed below:

Murata Gyrostar EBNC o3J

external voltage	$[+2.7\text{V}, +5.5\text{V}]$
current	$5\,\text{mA}$
max. angular velocity	$\pm300°/\text{s}$
output at $\omega = 0$	$+1.35\,\text{V}$
scale factor	$0{,}67\,\frac{\text{mV}}{°/\text{s}}$
temperature coefficient	$\pm20\%$
linearity	$\pm0{,}5\%$ of maximum signal
dimensions	$15.5\,\text{mm}\times8.0\,\text{mm}\times4.3\,\text{mm}$
measuring rate	$50\,\text{Hz}$ maximum
weight	$1.0\,\text{g}$

3.4 Distance sensors

Distance or proximity sensors are crucial for autonomous vehicles regarding collision avoidance and mapping. Distance sensors can be active or passive. Besides passive camera systems, there are active systems based on measurement principles like ultrasonic, infrared or lasers, see figure 3.17.

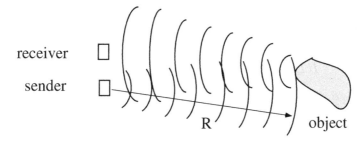

Figure 3.17 An active distance measuring system

For this kind of sensors the energy taken up at the site of the sensor depends on

- the sensing surface B of the receiver,

- the surface A of the object hit by the beam,

- the reflection $\rho(\alpha)$ of the beam from the object, being dependent on the angle of incidence to the surface normal and the reflectivity ($\geq 99\%$ for mirrors down to $0{,}01\%$ for strong absorbing surfaces like black soot).

The light intensity at the object $f(R)$ depends on the type of light source:

- $f(R) = \alpha = const$ for a laser beam hitting the object fully.

- $f(R) = a/R^2$ for other light sources with large radiation cones compared to the object,

- $f(R) = e^{-R/\gamma} \cdot a/R^2$ holds true if the light is being absorbed in the medium like e. g. light in fog or ultra sound in air.

At the object the reflected signal is $I_1 = A \cdot \rho(\alpha) \cdot I(R)$ with A the surface of the object for a normal source or surface of the laser beam with diameter Φ if the surface of the object is larger than $\pi/4 \cdot \Phi^2$.

$\rho(\alpha)$ is the reflectivity depending on impact angle α. At the site of the detector, the intensity is $I(D) = I_1 \cdot f_2(R) \cdot B$ with $f_2(R) = b/R^2$, b the proportion cut out of the reflected beam by the detector, and B the surface of the detector. Together this can be summarized to

$$I(D) = A \cdot I_0 \cdot \rho(\alpha) \cdot f(R) \cdot f_2(R) \cdot B \qquad (3.2)$$

$$\boxed{I(D) = A \cdot B \cdot 1/R^4 \cdot I_0 \cdot \rho(\alpha) \cdot a \cdot b} \qquad (3.3)$$

This is the **Radar equation** without a laser and without absorption. Note the strong dependency on the distance with the inverse fourth power in R. Using a laser beam with a surface area of $\pi/4 \cdot \Phi^2$ the radar equation becomes

$$I(D) = \pi/4 \cdot \Phi^2 \cdot B \cdot 1/R^2 \cdot I_0 \cdot \rho(\alpha) \cdot a \cdot b \qquad (3.4)$$

The advantage of using a sharply bundled illuminating source like a laser beam is the reduced dependency on R:
$I(D) \propto R^{-4} \longrightarrow I(D) \propto R^{-2}$.

In the following text different types of distance sensor for which the radar equation can be used are presented.

3.4.1   Infrared sensors

Distance measurement with infrared light is typically used as proximity sensors to detect nearby obstacles (5–80 cm).

In autonomous vehicles it is often applied as safety sensor, e. g. for the detection of a step or, in small robots, as a cheap sensor to explore the environment.

A triangulation sensor based on infrared was introduced by Sharp.[5] This sensor has a close-up range of 8–80 cm and delivers distance data with high precision and high resolution. In this sensor type the reflected light is projected onto a PSD or a 1D CCD-camera. The position x of the reflected light on the line camera with distance l to the infrared sender can directly be converted into the distance d to an obstacle with $d = f \cdot l/x$ (f is the focal length). Figure 3.18 shows the principle. The angular shift of ϵ between the focal plane of the lens and the CCD-line corrects for the shift in focal length for different distances. Figure 3.19 shows the principle of two reflective sensors for small distances.

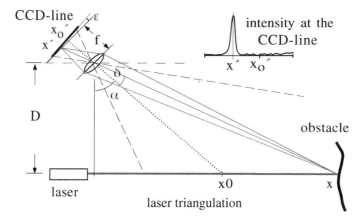

Figure 3.18 Measurement principle of an infrared triangulation sensor. x is the distance, f the focal length, D the distance between sender and linear camera and x' the intensity maximum in the linear camera picture.

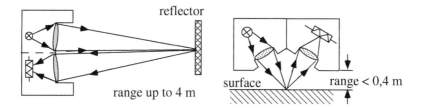

Figure 3.19 Reflexion light sensors; larger and short range

---

[5] http://www.sharpsme.com/Page.aspx/europe/en/

Figure 3.20 This forklift robot uses infrared sensors for collision avoidance.

3.4.2   Ultrasonic sensors

An ultrasound sensor bounces off a broad cone of ultrasound and measures an echo signal from obstacles in this cone. Ultrasonic sensors are often used as safety sensors in autonomous vehicles. To create ultrasound waves two principals are used: oscillating membranes and piezo crystals.

A thin metallic membrane of 50 mm diameter is electrostatically set into oscillations at a frequency around 50 kHz. The membrane sends out ultrasound pressure waves at a velocity of $c_{us} = 330$ m/s towards an obstacle. A reflected signal is picked up by the same membrane that is now used as a receiver. The time-of-flight is a measure for the distance to an object. In contrast to audible sound, ultrasound signals are attenuated severely in air, 11 m are the limit for 50 kHz signals. From the time t_L between start and first echo the distance is calculated as $L = c_{us}t_L/2$. The velocity of sound changes with temperature, air pressure and moisture content but may be treated as constant for the small distances of interest here. At 50 kHz the wavelength is $\lambda = 6{,}6$ mm.

The other type of ultrasound sensor is based on a piezo crystal. A piezo crystal changes its width if a voltage is applied to both sides of a crystal plate or produces a voltage under compression. A Sick sensor[6] presented in figure 3.23 uses this principle.

---

[6] http://www.sick.de/products/categories/industrial/ultrasonic/de/html

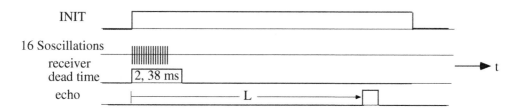

Figure 3.21 The operating principle of a emitter-receiver-module (Transducer) developed by Polaroid. After an initialization phase the polaroid sensor, with a membrane diameter 50 mm and a frequency of $f = 49.1$ kHz, sends out 16 oscillations in a cone angle of $30°$. The time to the return of the reflected signal is measured. The measure range of the sensor is 20 cm–10.5 m with a resolution of 1%.

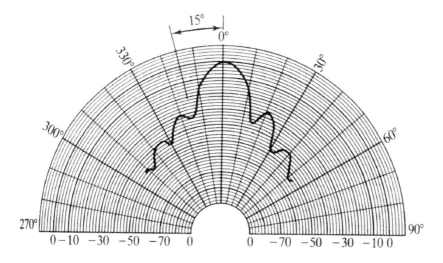

Figure 3.22 Cone of an ultrasound transceiver

Figure 3.23 Ultrasound sensor

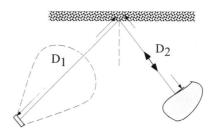

Figure 3.24 Hard surface reflecting like a mirror

Ultrasonic distance measurement also has disadvantages:

Surface roughness Any surface with a roughness that is small compared to the ultrasound wavelength used acts as a mirror, deflecting the sound without a return signal. For 50 kHz ultrasound any wall with surface roughness less than 5 mm behaves this way.

Multiple Reflections Misreadings are possible if the reflected signal hits an obstacle before it is bounced back to the sensor. In this case a distance much larger than the real one is calculated (in figure 3.24 $D_1 + D_2$ instead of D_1).

Soft materials Soft materials like pillows or curtains are strong ultrasound absorbers and can not be seen by the sensor.

Similar external signal An external ultrasound signal like pressurized air leaking into the environment could be misinterpreted if it has the same frequency range as the ultra sound sensor.

Crosstalk If more than one sender is firing simultaneously, the returning signals might overlap. This is called crosstalk.

3.4.3   Correlation of ultrasound signals

Some of the deficiencies occurring with ultrasonic measurements can be prevented by introducing correlation techniques [Joe98]. How similar are two given signals $x(t)$ and $y(t)$ to each other? The idea is to multiply the signals with a built-in time shift τ in one signal and integrate the product. The time shift is the characterizing parameter.

Crosscorrelation

$$P_{xy}(\tau) = \frac{1}{2T} \int_{-T}^{+T} x(t)y(t-\tau)dt \tag{3.5}$$

We see that a strong peak will occur whenever the signals are similar if shifted by a time τ_0 as shown in figure 3.25.

Autocorrelation

$$P_{xx}(\tau) = \frac{1}{2T} \int_{-T}^{+T} x(t)x(t-\tau)dt \tag{3.6}$$

This causes a peak whenever there are periods in a signal. For uncorrelated signals there is a single peak at $\tau = 0$, as shown in figure 3.26. Signals in the form of broadband noise have small crosscorrelation and good autocorrelation, as shown in figure 3.27.

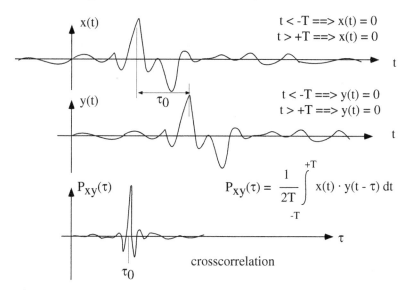

Figure 3.25 Crosscorrelation

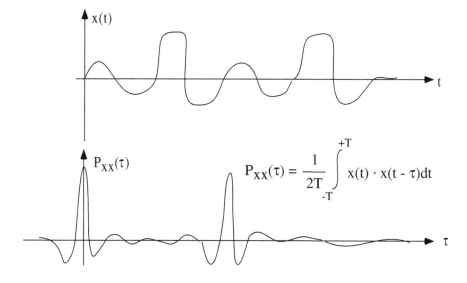

Figure 3.26 Autocorrelation

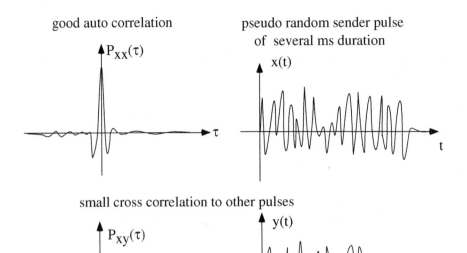

Figure 3.27 Pseudo random pulse

The correlation of signals $x(t)$ and $y(t - \tau)$ may be calculated using Fourier transforms:

$$P_{xy}(\tau) = \frac{1}{2\pi} \int_{-\infty}^{+\infty} F^*(\omega) \cdot G(\omega) e^{-i\omega\tau} d\tau \qquad (3.7)$$

$$F^*(\omega) = \int_{-\infty}^{\infty} x(t) \cdot e^{-i\omega t} dt \qquad (3.8)$$

$$G(\omega) = \int_{-\infty}^{+\infty} y(t - \tau) e^{i\omega t} dt \qquad (3.9)$$

Let $x(t) = 0$ and $y(t) = 0$ for $t < 0$ and $t \geq N \cdot \Delta t$ denote two signals. Discretizing $x(t) \longrightarrow [x_j]$ and $y(t) \longrightarrow [y_j]$ under the discretization $x_j = x(j \cdot \Delta t)$ allows us to calculate the Fast Fourier Transforms (FFT) $F(\omega) \longrightarrow [F_k]$ and $G(\omega) \longrightarrow [G_k]$.

$$F_k^* = \sum_{j=0}^{N-1} x_j \cdot e^{-i2\pi jk/N} \qquad (3.10)$$

$$G_k = \sum_{j=0}^{N-1} y_j \cdot e^{i2\pi jk/N} \qquad (3.11)$$

where $j,k = 0,\dots,N-1$ and $N = 2^n$. The back transformation delivers $P_{xy}(\tau) \longrightarrow [P_{xy}(\tau_m)]$ with $\tau_m = m \cdot \Delta t$

$$P_{xy}(\tau_m) = \frac{1}{2\pi} \sum_{j=0}^{N-1} F_k^* \cdot G_k e^{-i2\pi mk/N} \tag{3.12}$$

Using suitable pseudo random signals of 10 ms duration correlating the transmitted and received impulses allows a precise measurement of distances despite crosstalk. Two objects nearby each other may be discriminated as shown in figure 3.28.

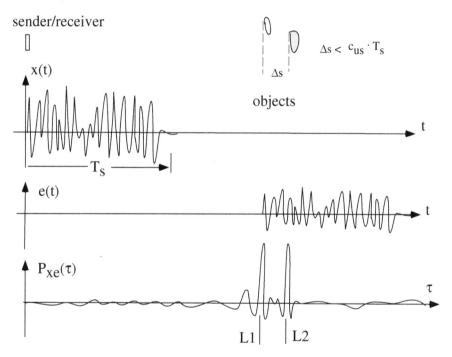

Figure 3.28 Better distance resolution

3.4.4 Laser sensors

Laser radar sensor The most prominent sensor of this type is the Sick PLS sensor.[7] It is an active laser sensor operating on narrow pulses of 1 ns length but 10 W power with 4500 pulses/s (eye safe) in the near infrared section of

---

[7] Sick Optoelectronic Waldkirch, Baden, Germany; http://www.sick.de/de/products/categories/safety/de.html

the spectrum. On emission of a pulse, a fast counter is started and stopped again at the first return of a signal, picked up by a rather large mirror and an avalanche photo diode. The counter directly measures the distance to an obstacle. Within 1 ns the light travels 15 cm from the sender towards an obstacle and the same distance back again towards the receiver. Figure 3.29 illustrates this principle.

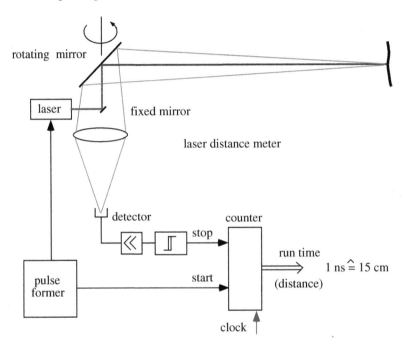

Figure 3.29 Principle of the Sick laser scanner

Each tick of the counter denotes a distance of 5 cm. As the emitted signal is rather strong, every surface with a reflectivity larger than 1.8% (black leather, dust on a polished glass surface or a fly in the beam) gives enough reflection to see a return signal and detect an obstacle. The scanning range of the sensor is 180°. 361 measured distances taken every 0,5° form a laser scan (running period 80 ms). Further information is printed in the table below. The sensor is licensed in the EU to be used to trigger an emergency stop for distances less than 4 m. Figure 3.30 shows a picture of the sensor.[8] Figure 3.31 shows a laser radar scan of a room with some furniture in it; the grid width is 1 m.

---

[8] http://www.sick.de/

The main parameters of a Sick laser range sensor are listed below:

Cone angle	180°
Angle resolution	1°/**0.5°** / 0.25°
Response time	13 ms / **26 ms** / 53 ms
Resolution	10 mm
Systematic error	±15 mm
Statical error at 1 σ	5 mm
Measurement range	8 m, 16 m, 32 m, **80 m**
Transfer rate	9.6/19.2/38.4/**500** kBaud
Working temperature	0...+50°C
Supply voltage	24 V ±15%
Weight	ca. 4.5 kg
Size (H × L × W)	210 mm × 156 mm × 155 mm

Figure 3.30 The Sick laser scanner LMS200

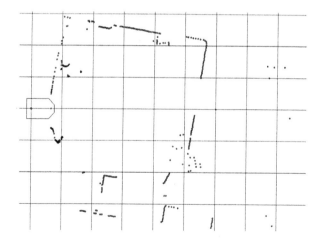

Figure 3.31 Typical laser scan of a room

3.5 Vision sensors

There are two important semiconductor sensors available for computer vision: CCD and CMOS. The basis of semiconductor cameras is the "inner photoelectric effect": In certain materials electrons are set free under photon absorption (light). The amount of electrons is correlated to the amount of photons. While CMOS sensors offer small size, bigger dynamics and show no blooming effects, the CCD sensor has higher photo sensitivity, higher uniformity and less noise.

3.5.1 CCD camera

CCD chips are sensitive to the complete visible spectrum, but especially to red light. Color images are created through application of RGB filters and a combination of 4 photo diodes for 1 pixel. One problem with CCD technology is the evaluation per column. Under intense light this leads to the so-called "blooming" where a column seems to be fully illuminated. During integration of charges the pool may be flooded. This effect can be avoided with "drain canals" on the chip, reduction of exposure time or reduction of shutter opening time.

3.5.2 CMOS camera

CMOS technology offers a continuous conversion of the photon beam into output voltage while each cell or pixel can be accessed directly. Therefore, CMOS is more expensive than CCD.

3.5.3 Stereo-camera systems

Binocular stereo vision Humans are able to perceive depth information with their two eyes. A distant object is projected onto the retina of the left and right eye. The object's projections differ in their position. The human brain is able to generate a depth judgment for that object by analyzing the two images.

Computer stereo vision copies this concept. The goal is to reconstruct a depth map from at least two 2D camera images showing a 3D scene from different observation points. The depth information can be inferred by matching a point in both images and looking at the displacement between the matched pair. Figure 3.32 visualizes the general case of stereo vision geometry with two input images.

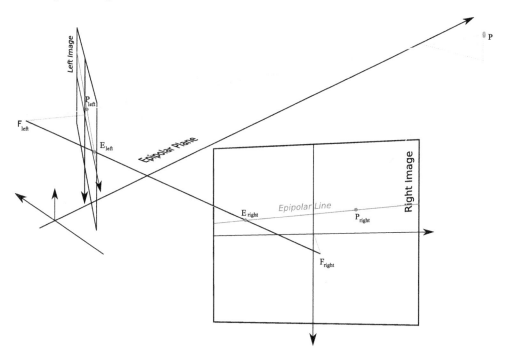

Figure 3.32 General stereo vision geometry (pin-hole camera model). E_{left}, E_{right}: epipolar points; F_{left}, F_{right}: focal points; P: point in 3D space; P_{left}, P_{right}: projections on image planes

A 3D scene is projected onto two 2D virtual image planes. The plane defined by point P and both focal points F_{left} and F_{right} is called *epipolar plane*.

The projected line from focal point F_{left} to P in the right image is called an *epipolar line*. The right image projection P_{right} of point P can always be found on this epipolar line. This is called the *epipolar constraint* and simplifies the correspondence problem. This constraint only holds for the perfectly rectified images of a pin-hole camera. In reality, raw camera images are usually distorted. The images have to be rectified prior to stereo vision processing.

A binocular stereo vision system with *canonical stereo geometry* consists of two identical cameras mounted in parallel. The cameras are placed so that their virtual image planes and both epipolar lines fall together (see figure 3.33). The epipolar lines are aligned parallel to the x-axis of the image planes.

Each point in the observed scene will be projected to the same row in the left and right image but on different pixels here. The displacement between left and right projection is called *disparity d*.

In order to reconstruct the 3D scene from disparity values, the fixed distance between the cameras called *baseline b* and the focal length f of the cameras has to be known.

The stereo vision head coordinate system lies between both cameras on the baseline. Its orientation is defined according to common decisions in robotics as a right-hand coordinate system. The x-axis pierces perpendicular through the image plane (see figure 3.33).

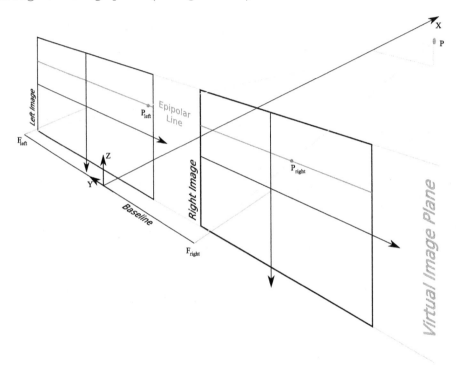

Figure 3.33 Binocular stereo vision setup

In computer vision, the origin of an image usually lies in its top left corner. Here, the image coordinate systems have their origins translated to the very middle of each picture, closest to the focal point (see figure 3.33). This is not a restriction. It only simplifies the following derivation and is a common assumption in physics for the pin-hole camera model.

The exact 3D coordinates x, y and z for two matching pixel (x_l, y_l) and (x_r, y_r) can be calculated using triangulation as shown in figure 3.34.

$$\tan(\alpha_l) = \frac{x_l}{f} = \frac{\frac{b}{2} - y}{x} \tag{3.13}$$

$$\tan(\alpha_r) = \frac{-x_r}{f} = \frac{\frac{b}{2} + y}{x} \tag{3.14}$$

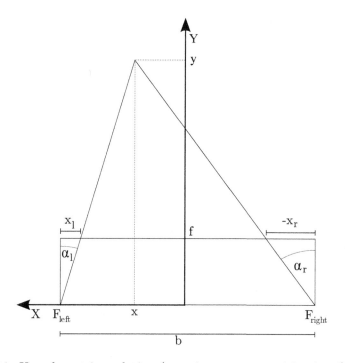

Figure 3.34 X-y-plane triangulation (top view on stereo vision head)

Solving 3.13 and 3.14 for y yields:

$$y = \frac{-x_l \cdot x}{f} + \frac{b}{2} \tag{3.15}$$

$$y = \frac{-x_r \cdot x}{f} - \frac{b}{2} \tag{3.16}$$

Equating 3.15 and 3.16 and solving for x results in:

$$x = \frac{b \cdot f}{x_l - x_r} \tag{3.17}$$

Solving for z is even simpler. Since both pixels lie on the same image row $y_r = y_l$ holds, see figure 3.35.

$$\tan(\beta) = \frac{y_l}{f} = \frac{z}{x} \tag{3.18}$$

$$z = \frac{x \cdot y_l}{f} \tag{3.19}$$

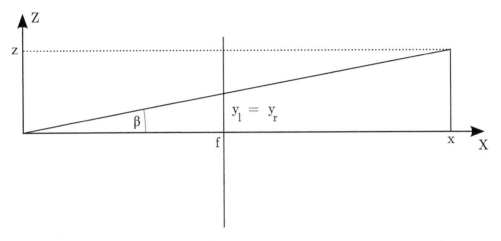

Figure 3.35 X-z-plane triangulation (side view on stereo vision head)

Substitution of equation 3.17 into 3.19 and 3.15 yields the following formulas for straight forward computation of y and z.

$$y = \frac{-\frac{b}{2} \cdot (x_l + x_r)}{x_l - x_r} \tag{3.20}$$

$$z = \frac{y_l \cdot b}{x_l - x_r} \tag{3.21}$$

The disparity value d calculated by a stereo vision algorithm already equals $x_l - x_r$. Note that the translation of the image origin along the x-axis has no effect on the disparity. Let the left image be the reference image. Then x_l and d are known. x_r can easily be calculated as $x_r = x_l - d$.

Stereovision algorithms The challenge for a stereo vision algorithm is to solve the correspondence problem, i. e. finding the right match in the left and right image for a given point in the real world. The common approach is to declare one input image as the reference image. For a pixel in the reference image the corresponding pixel is then searched for in the second input image.

However, finding the correct match is ambiguous. Due to occlusion some points may not have a match in the other projection (see figure 3.36). Furthermore, repetitive textures may produce multiple matches. Detecting occlusion and finding the right match determines the quality of a stereo vision algorithm.

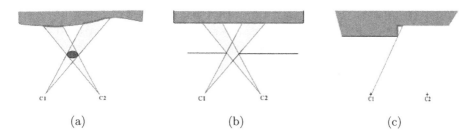

(a)   (b)   (c)

Figure 3.36 Scene configurations with half-occluded regions (highlighted in red): (a) occlusions due to thin object at the foreground-scene discontinuity, (b) occlusion due to a small hole at the foreground-scene discontinuity, (c) occlusion due to surface variation-surface discontinuity (source [Kos02]).

Stereo vision strategies can be dense or sparse. Sparse stereo vision concentrates on selected feature points only. A feature point can be a point of special interest. Alternatively, a feature point may be a point with high likelihood for a good match. This approach can potentially result in a fast stereo vision algorithm due to the reduced amount of analyzed points. Dense stereo vision algorithms find matches for all points in the reference image (a dense approach was chosen here since distance information for the whole image is required for further processing steps). In a complete run of a dense stereo vision algorithm, a disparity value is calculated for each pixel in the reference image. Typically this result is visualized as a gray scale image, also known as disparity map. In this image each pixel's brightness corresponds to a disparity level. High disparity values result in lighter pixels.

Instead of finding suitable feature points in a pre-processing step, the disparities for all pixels are computed in parallel. Therefore, points with high confidence can serve as sparse feature points if needed.

Global vs. window-based optimization   Recently, high quality results have been achieved in CPU stereo vision by applying global optimization techniques to the stereo vision problem (see [SS02]). Such algorithms rank top at the Middlebury evaluation page.[9] However, these approaches tend to be slow and thus are not suitable for our near-real-time task. Yang and Pollefeys [YP05] claim that only correlation-based stereo algorithms can provide

---

[9] http://vision.middlebury.edu/stereo/

a dense depth map in real time on standard computer hardware. However, a fast optimizing stereo vision algorithm for the GPU could be an interesting future research topic.

GPU computation profits from parallelizing, while inter-process communication is extremely limited[10] or expensive[11]. In this work a window-based approach was chosen.

Window-based stereo algorithms take two parameters: window width and window height. The window describes a rectangle area around the reference pixel. Around a match candidate in the second image a rectangle of equal size is compared to the reference window, see figure 3.37. The most similar areas are taken to be the correct match.

Figure 3.37 Window-based stereo vision

For each disparity step a window-based strategy shifts the window along the epipolar line in the second image. The region in both images is then compared by a similarity function. The disparity step with the highest similarity most likely provides the correct match. Common similarity functions are cross correlation, sum of squared differences (SSD) or sum of absolute differences (SAD).

The performance and quality of a window-based algorithm depends on the window size. Large support regions provide a higher certainty for a correct match even for low textured input images. Small window sizes speed up the computation and produce finer grained results. Window-based stereo vision requires good textures in the input images. Luckily, in the application field of outdoor robotics such textures prevail.

---

[10] shared memory accessible for threads in same block only
[11] global memory access is slow

4 Localization

For autonomous navigation, a mobile robot needs to consider its position and orientation within a certain working coordinate frame. The so-called localization problem is usually solved by a mixture of different principles contributing to the state variables of position and orientation, called "pose" in the following. Generally two different approaches are being considered: absolute pose determination and relative pose determination. For each time step, the incremental observation of velocities, angular velocities, forces, or visual characteristics give an information update for the pose variables. Absolute information derived from global landmarks like GPS, however, directly provides position information. In the following, the important variants of pose measurement as well as the required principles of selected sensor systems are described. Also some exemplary implementations are explained in more detail. There are several possibilities to group the different techniques, such as regarding absolute or incremental information or position and orientation sensors. However, it seems most logical to present the measurement principles in an order they can be put together as a whole localization system, supplementing each other.

4.1 Pose calculation from odometry

A robot's trajectory derived from the summation of wheel velocities is called odometry. In order to exemplify the techniques involved, the calculations for a differential drive are presented in the following.

Let two wheels roll on the ground with distance d between them, see figure 4.1. While driving in a circle with a radius R, the driven distance of each wheel (s_1 and s_2) and the angle φ at which the vehicle turns measured by the difference in ticks of the right and the left wheel n_1 and n_2 are calculated as given in equations 4.1 to 4.3. Here n_0 denotes the number of ticks per wheel revolution and r denotes the wheel radius.

$$s_1 = 2\pi(R + d/2)\varphi/360° = n_1\frac{2\pi r}{n_0} \tag{4.1}$$

$$s_2 = 2\pi(R + d/2)\varphi/360° = n_2\frac{2\pi r}{n_0} \tag{4.2}$$

$$\varphi = \frac{360°(n_1 - n_2)r}{d \cdot n_0} \tag{4.3}$$

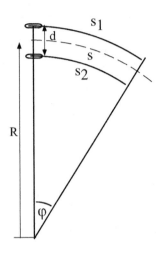

Figure 4.1 Measuring turning angles

Any small difference in the radii of the right and left wheel shows up in an angle indistinguishable from a real turning.

Let $r_2 = r$ and $r_1 = r + \delta r$. Then $\delta\varphi = 360° \cdot n_1/n_0 \cdot \delta r/d$. So after a while, the measurement significantly deviates from the actual rotation performed by the vehicle. As an example let $r = 100\,\text{mm}$, $\delta r = 1\,\text{mm}$, $s = 6\,\text{m}$ and $d = 500\,\text{mm}$. This leads to a deviation of $\delta\varphi = 7.2°$.

Measuring path lengths is a lot more precise as the weighted sum of the radii are taken:

$$s = \frac{s_1 + s_2}{2} = \frac{n_1 + n_2}{2}\frac{2\pi r}{n_0} \tag{4.4}$$

As already discussed in section 2.2.4, the position and orientation of a vehicle can be determined using equations 2.35 through 2.37. These integrals are generally not solvable in closed form. If $v(t)$ and $\omega(t)$ are measured every Δt, the integrals have to be calculated from sets $[v_i]$ and $[\omega_i]$. The time interval Δt must be short enough to allow the proper reconstruction of $\psi(t)$ according to figure 4.2. The numerical integration, using the simple trapezoid formula, gives the formulas 4.5 to 4.7.

$$\psi_i \approx \psi_{i-1} + (\omega_{i-1} + \omega_i) \cdot \Delta t/2 \tag{4.5}$$
$$x_i \approx x_{i-1} + (v_{i-1}\cos(\psi_{i-1}) + v_i\cos(\psi_i)) \cdot \Delta t/2 \tag{4.6}$$
$$y_i \approx y_{i-1} + (v_{i-1}\sin(\psi_{i-1}) + v_i\sin(\psi_i)) \cdot \Delta t/2 \tag{4.7}$$

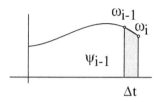

Figure 4.2 Principle of numerical integration

While the distance $s(t)$ driven can be measured by internal odometry rather precisely to better than 1%, the calculation of $x(t)$ and $y(t)$ requires the knowledge of the angle $\psi(t)$. Using wheel encoders only, the differences in wheel radii sum up integrating $\omega(t)$, while in the measured distance only the weighted sums of the radii are integrated. Figure 4.3 shows the error summing up. Using only wheel encoder data typical error ranges for ψ are approximately 1°/m, i. e. 0.174 m deviation to the side over a path length of 10 m.

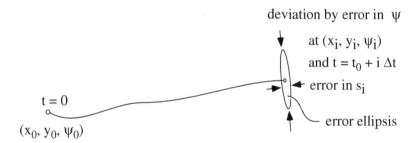

Figure 4.3 Position error ellipsis

As the error in s is small and the error in ψ large in comparison, this results in an elongated error ellipsis. The error ellipsis describes the distribution of measured points around the calculated point (x_i, y_i). Under the assumption of equal distributed errors, the probability of finding a distance s although the distance s_i has been travelled, is

$$p(s) = \frac{1}{\sqrt{2}\sigma_s} \exp\left(-\frac{(s-si)^2}{2\sigma_s^2}\right) \tag{4.8}$$

This is a Gauss normal distribution. In figure 4.3 the width of the ellipsis is just σ_s. By the same argument the distribution in ψ is given by

$$p(\psi) = \frac{1}{\sqrt{2}\sigma_\psi} \exp\left(-\frac{(\psi - \psi_i)^2}{2\sigma_\psi^2}\right) \tag{4.9}$$

The general problem of odometry in its incremental nature is that real systems show no longterm stability. Slippage, measurement errors, and in the case of 3D (outdoor) environments one missing measurement direction, lead to the problem of the calculated pose information going wrong after a certain amount of time. Odometry has therefore to be combined with information from absolute pose measuring sensor systems.

4.2 Inertial measurement units (IMU)

The name inertial measurement unit, also called inertial system, derives from the fact that only inner information of a system, forces and angular velocities (in three dimensions in this case) are regarded and therefore the system is not vulnerable to errors like slippage that influence for example the odometry calculation. The important components of such a system are depicted in figure 4.4. Physics teaches us that each change in direction of a movement manifests itself in a measurable acceleration force. Being in free space, a moving system therefore needs to be accelerated to change direction and/or velocity. Therefore, measuring acceleration forces allows us to reason about state changes in this respect and to interpret the pose change occurring, with given knowledge about the initial unaccelerated state, possibly including a fixed velocity component.

Two important aspects are mentioned in advance. Firstly, we are not in free space, therefore the earth's rotation and gravity force lead to some more or less disastrous impact on our measurements. Secondly, the pose information derived is, as with odometry, of incremental nature. However, the example described in section 4.2.2 contains some heuristics that enhance inertial systems to work at least as an absolute sensor for orientation, turning the necessity to deal with gravity force into a virtue. The general algorithm for calculation of pose information from inertial measurement considering all effects on or in proximity of our planet is depicted in figure 4.5. The Coriolis force, earth rotation etc. have to be considered, for example when using inertial systems as navigation support in planes. Inertial measurement units have been used on autonomous robots for a long time, see [BF94, NDW99,

HKNO01]. However, in mobile robotics a simplified approach can be used as described in section 4.2.2.

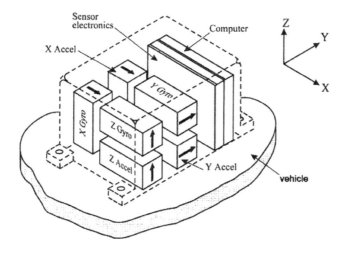

Figure 4.4 Components of an inertial measurement unit

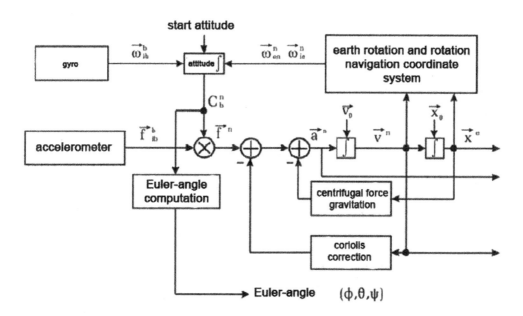

Figure 4.5 General algorithm for inertial pose calculation

4.2.1 Simplified inertial calculation

As described at the beginning of this section, the pose information is derived using sensors for angular velocity $\omega_{x,y,z}$ and linear acceleration $a_{x,y,z}$. The following notations are used:

Inertial calculation expressions:
$\vec{\omega}$: measured angular velocities
$\vec{\sigma}$: angles from integration of angular velocity
ϕ: roll-angle around x-axis
θ: pitch-angle around y-axis
ψ: yaw-angle around z-axis
$\vec{a}$: accelerations
$\vec{g}$: gravity force
$\vec{v}$: velocities
$\vec{s}$: position
$\vec{u}$: rotation axis for attitude correction heuristic
α: rotation angle for attitude correction heuristic
t: time interval
W: matrix with unit vectors of global frame
B: matrix with unit vectors of local frame
T: rotation tensor, translates from local to global frame
ε: tolerance interval
lower index b: vector/matrix referred to local frame
lower index w: vector/matrix referred to global frame
double lower index $\vec{\alpha},\beta$: tensor rotates around axis $\vec{\alpha}$ by angle β
upper index k: vector/matrix/tensor in kth time step

The world model shall be an unaccelerated plane surface only interfered with by the gravity force. The matrix B_w with axis vectors of the body frame from world view contains the attitude information of the IMU. At this part the common strap-down approach [TW97] is modified, where the attitude information is only stored in three angle variables. Using this axis matrix B_w has the advantage that the attitude is well known and a calculation of angles necessary for an representation by individual rotations can be done if necessary. The matrix, however, is updated in every step of the calculation.

$$W_w = \begin{pmatrix} 1 & 0 & 0 \\ 0 & 1 & 0 \\ 0 & 0 & 1 \end{pmatrix} \tag{4.10}$$

$$B_w = \begin{pmatrix} b_{11} & b_{12} & b_{13} \\ b_{21} & b_{22} & b_{23} \\ b_{31} & b_{32} & b_{33} \end{pmatrix} \tag{4.11}$$

The angle change per time step is deduced by trapezoid integration of the current and last angular velocity measurement. This angle needs to be transformed to the global frame.

$$\vec{\sigma}_b^k = t^k \left(\vec{\omega}_b^k + \vec{\omega}^{k-1} \right) / 2 \tag{4.12}$$

$$\vec{\sigma}_w^k = B_w^k \vec{\sigma}_b^k \tag{4.13}$$

Attitude information is updated by multiplication of B_w with a rotation tensor. This is possible as this matrix contains the axis vectors of the body frame. These vectors are rotated at this point. The rotation axis represented by the tensor equals the angular rate. The rotation angle is derived as the length of the current angular velocity vector. Only two of three normal vectors (rows in the orientation matrix B_w) are rotated variantly. The third one is calculated by cross product. On one hand, this saves calculating time and on the other hand, it guarantees the rectangularity and normalization of the unit vectors in the transformation matrix.

$$B_w^{k+1} = T_{\vec{\sigma}^k, |\vec{\sigma}^k|} \cdot B_w^k \tag{4.14}$$

The matrix which transforms from the local coordinate frame to the global frame is conveniently the attitude representation itself, as shown below. D shall be the initially unknown transformation matrix here.

$$B_w = DW_w \tag{4.15}$$

$$W_w = I \tag{4.16}$$

$$B_w = D \tag{4.17}$$

For sensor fusion and graphical output, the attitude needs to be represented by Euler angles. The combined Euler matrix represents three sequential rotations by the angles ϕ, θ and ψ around the world axes x, y, and z, or the body axes in opposite order (c: cos, s: sin):

$$D = \begin{pmatrix} c\psi c\theta & -s\psi c\phi + c\psi s\theta s\phi & s\psi s\phi + c\psi s\theta c\phi \\ s\psi c\theta & c\psi c\phi + s\psi s\theta s\phi & -c\psi s\phi + s\psi s\theta c\phi \\ -s\theta & c\theta s\phi & c\theta c\phi \end{pmatrix} \tag{4.18}$$

The angles can be determined by coefficient comparison from B_w.

$$\phi = \arctan\left(\frac{b_{32}}{b_{33}}\right) \tag{4.19}$$

$$\theta = \arcsin\left(-b_{31}\right) \tag{4.20}$$

$$\psi = \arctan\left(\frac{b_{21}}{b_{11}}\right) \tag{4.21}$$

Knowing the orientation, the measured acceleration forces can be translated. By multiplication the matrix B_w, they are mapped to the global frame and are used to continuously integrate the acceleration of the system to get velocity and position. The force of gravity has to be subtracted in every time step. Here we have the largest source of position errors. A wrong attitude causes a wrong difference and an incorrect acceleration vector is processed.

$$\vec{g}_w = \begin{pmatrix} 0 \\ 0 \\ 9{,}80665\,\text{m/s}^2 \end{pmatrix} \tag{4.22}$$

$$\vec{a}_w = B_w\vec{a}_b - \vec{g}_w \tag{4.23}$$

$$\vec{v}_w^k = \vec{v}_w^{k-1} + t^k\left(\vec{a}_w^k + \vec{a}_w^{k-1}\right)/2 \tag{4.24}$$

$$\vec{s}_w^k = \vec{s}_w^{k-1} + t^k\left(\vec{v}_w^k + \vec{v}_w^{k-1}\right)/2 \tag{4.25}$$

Unfortunately, a number of errors is totalized due to integration and lead to an unbearable pose error in a rather short time. The errors which can be identified are assembly errors (alignment on board, rectangularity), electrical errors (voltage fluctuations, system noise, digital jitter), and sensor errors (temperature dependence, cross axis influence, characteristic curve).

4.2.2 Implementation example with heuristics for absolute orientation measurement

The inertial system shown in figure 4.6 was developed at the Robotics Research Lab and is currently used primarily as orientation sensor for the autonomous outdoor robot RAVON. One major aspect for the success of the system as a sensor on RAVON is the absolute orientation information that improves calculation of trajectory from odometry. As stated already, the odometry consists of a scalar velocity information of each wheel and therefore gives only limited information about angular velocity which, however, is crucial for localization. Therefore, a 3D orientation sensor adds the missing information to the incremental information from the wheels. However, without global positioning with visual landmarks or GPS, the combined system would still lack long term-stability. A more detailed description of the system used in this localization approach can be found in [KHB05].

Figure 4.6 Inertial system developed at the RRLab

Gravity correction heuristic The absolute orientation information mentioned above is derived via a heuristic depending on knowledge about the earth's gravity force. A component for inertial gravity correction provides the orientation towards the earth's surface φ, θ by comparison with an estimated gravity vector.

In order to minimize drift in position and attitude, one possibility is to adjust the attitude with respect to the gravity force. In each calculation step, an attitude information is determined. If the total of the acceleration vector equals the standard gravity force, a first criterion is fulfilled.

$$||\vec{a_w}| - |\vec{g}|| \leq \varepsilon \qquad (4.26)$$

If it differs in direction and not in length from the current acceleration vector, a correction of the attitude representation is done by rotating it so that the two vectors point in the same direction. The rotation is done respective to the axis built from the cross product of the two vectors while the rotation angle arises from their different directions.

$$\alpha = \arccos \left(\frac{\vec{a_w}\vec{g}}{|\vec{a_w}||\vec{g}|} \right) \qquad (4.27)$$

As a result of the lack of one dimension, only the roll and pitch values can be corrected by this method.

$$\vec{u} = \vec{a_w} \times \vec{g} \tag{4.28}$$

$$B_w = T_{\vec{u},\alpha} \cdot B_w \tag{4.29}$$

This attitude correction is only heuristic and therefore is only applied if the deviation reaches a certain amount. Additionally, the correction is suppressed while the acceleration data is fluctuating. However, the acceleration vector might have the same length as the gravity vector. Nevertheless, in all experiments the correction helped greatly to improve the overall performance.

Calibration To get the best possible results despite of the listed error sources, a system should be calibrated thoroughly. The measured value $\tilde{x}$ can be described as polynomial function of the true value x and the error coefficients e_i.

$$\tilde{x} = e_n x^n + e_{n-1} x^{n-1} + \cdots + e_1 x + e_0$$
$$= \sum_{n=0}^{n} e_n x^n \tag{4.30}$$

The coefficients of a suitable linear error model can be determined by taking representative measurements. This model consist of offsets, linear scale factors and a cross correlation matrix. These are 12 unknown variables for the set of angular velocity sensors and another 12 for the set of acceleration sensors if we move the scale factors to the diagonal elements of the correlation matrix. The unknown coefficients could be determined by taking a couple of linear independent measurements and solving an equation system, but indeed a gradient descent procedure shows better results. The test measurements are compared to their goal results. An error sum represents a simple fitness function. The gradient descent is done by starting with plausible initial values for the unknown variables and making random changes trying to minimize the differences between measured test vectors and their corresponding goals. If the error decreases in one loop, the achieved set of values becomes the basis for the next step. This strategy leads to an optimal solution as the model is linear.

$$\begin{pmatrix} x_1 \\ x_2 \\ x_3 \end{pmatrix} = \begin{pmatrix} k_{11} & k_{12} & k_{13} \\ k_{21} & k_{22} & k_{23} \\ k_{31} & k_{32} & k_{33} \end{pmatrix} \cdot \left[\begin{pmatrix} a_1 \\ a_2 \\ a_3 \end{pmatrix} + \begin{pmatrix} \tilde{x}_1 \\ \tilde{x}_2 \\ \tilde{x}_3 \end{pmatrix} \right] \tag{4.31}$$

The output of the angular velocity sensors depends on their temperature. Therefore, at least the offset has to be gathered at different temperatures to make linear interpolation possible.

4.3 Localization based on optical flow

The idea of visual odometry was first developed by L. Matthies [Mat89]. His approach was refined and deployed on several Mars robots [CMM06] and other researchers implemented different flavors to the original concept. The approaches mostly differ in the camera system which is the basis for computing visual odometry. In [CSNP05] a monocular, in [CSS04] an omni-directional and in [Mat89] a stereo system is used. In principle most visual odometry approaches follows the pattern outlined in figure 4.7.

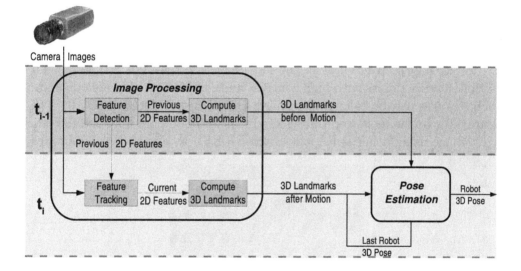

Figure 4.7 Overview of visual odometry

The camera system captures a frame, and a set of features suitable for tracking is selected. After a while, another frame is captured. The features that have been detected in the previous frame are now tracked into the current frame. The vectors between these points contain the information about the change in orientation and translation of the robot between the two subsequent frames. The last step is an ego-motion estimation on the basis of the vectors to receive the robot pose according to a given starting point.

In the following a visual odometry approach based on stereo-vision is described. This concept, shown in figure 4.8, is deployed on the outdoor robot RAVON of the University of Kaiserslautern.

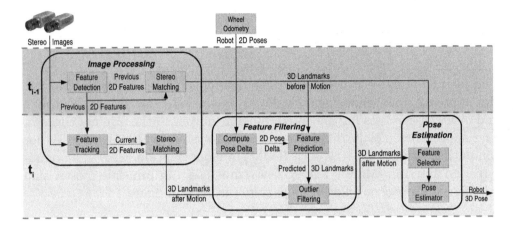

Figure 4.8 Concept of visual odometry for the outdoor robot RAVON

In the first step features in the left image are selected. For stereo-ego-motion, feature points with high contrast and minimal spacing are adequate. Such features can be calculated using a Harris corner detector [Der04] (see figure 4.9). In the next step the selected feature points are matched into the right image using a Lucas-Kanade-tracker [LK81] to obtain 3D coordinates.

Figure 4.9 Feature selection

The third step is the tracking of the selected features into the subsequent left image of the stereo pair, see figure 4.10. After that, another stereo matching step is carried out. The above procedure yields pairs of consecutive 3D feature locations. These can now be used to determine the translation vector (T) and rotation matrix (R) which describe the movement of the robot between the two frames in question.

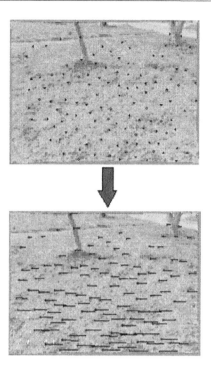

Figure 4.10 Feature tracking

Equation 4.32 shows the mathematical connection between robot pose transition and the tracked features where L_i^a are the features at the previous and L_i^b the features at current robot position.

$$L_i^b = RL_i^a + T + e_i \qquad (4.32)$$

e_i denotes the position error. To obtain an optimal motion estimation this error has to be minimal for all pairs of features. To solve this problem, a least square optimization is carried out (see equation 4.33).

$$e^2 = \frac{1}{N} \sum_{i=1}^{n} \| L - i^b - (RL_i^a + T) \| \qquad (4.33)$$

The visual odometry calculation was applied in a realistic simulation of RAVON in a typical test scenario. The driven track has a length of 21 m. The endpoint deviation was about 1.75% (0.38 m) and the mean deviation 0.15 m. Because of the large amount of disturbances when driving in outdoor terrain, the mean error increases for a track of 100 m to about 10%. One of the main problems of error rise are dynamical objects like cars, people or moving plants, because it has to be decided if the movement of the features is based on robot motion or on object movement.

To eliminate this error, odometry could be used as a filter. Therefore, the difference of captured and predicted 3D position (using pose estimation of the odometry) of the feature points is computed in each step. All differences above a specific threshold indicate that with a high probability features of a dynamical object have been tracked. These features are not used to calculate the new position of the vehicle.

4.4 Feature extraction from laser radar data

Distance sensors, notably laser range sensors, give a set of ordered distance points (r_i, ϕ_i) with respect to the vehicle coordinate system. From this cloud of points, information about the environment of the vehicle can be obtained.

4.4.1 Obstacles

A simple information from a laser radar set regards obstacles in the vicinity of the vehicle. Any point in a cone of angle α with distances less than r_0 is an obstacle to be avoided. Given the momentary direction, the free zone depends on the dimensions of the vehicle. There are two critical distances: r_{02}, where the vehicle control can still react to find a path around the obstacle, and r_{01}, where an emergency break has to be raised. Let b be the width of the vehicle. Therefore, the cone angle is given by $\alpha = 2 \arctan(b/2r_0)$. This defines a safety fence around the vehicle.

4.4.2 Line extraction

Any straight structure in the environment of the vehicle, indoors mainly straight walls, presents itselves as a group of distance points lying more or less in a line. To extract these lines a histogram may be used.

Histogram algorithm   A scan of the environment with distance points (r_i, ϕ_i) as shown in figure 4.11 and figure 4.12 is given. The angle ϕ_i modulo 180° of a line connecting a pair of distance points (r_i, ϕ_i) and (r_{i+k}, ϕ_{i+k}) and the vehicle axis can be calculated. k has to be chosen in a way such that the inevitable errors in distances are smoothed out. In many cases k=3 to 5 will do.

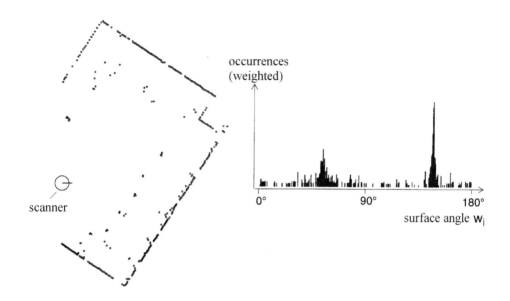

Figure 4.11 A real scene and its histogram

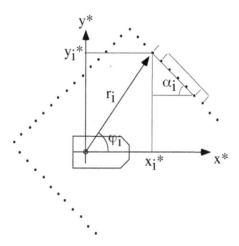

Figure 4.12 Determination of angles of walls

In a vehicle centered coordinate system a measured point (r_i, ϕ_i) has the coordinates

$$x_i = r_i \cos \phi_i \qquad (4.34)$$

$$y_i = r_i \sin \phi_i \qquad (4.35)$$

Then the angle α_i is given by

$$\tan \alpha_i = \frac{y_i - y_{i+k}}{x_i - x_{i+k}} \tag{4.36}$$

$$\tan \alpha_i = \frac{r_i \sin \phi_i - r_{i+k} \sin \phi_{i+k}}{r_i \cos \phi_i - r_{i+k} \cos \phi_{i+k}} \tag{4.37}$$

Afterwards a weight function is applied to the angle and it is sorted into a histogram with a box width $\delta\alpha$. The box width should preferably be the angular resolution of the distance sensor. The last step is to be repeated for all indices i. Once all points are processed, the number of weights in each box has to be determined. Figure 4.13 shows the distribution of angles of the point cloud with respect to the vehicle axis. In a rectangular room there are two peaks in the histogram $90°$ apart. Subsequently the center of gravity as the weighted sum over the number n_j in box $j\delta\alpha$ for all j around a peak in the distribution has to be determined.

$$\Phi = \sum_{j=j_1}^{j_2} n_j j \delta\alpha / \sum_{j=j_1}^{j_2} n_j \tag{4.38}$$

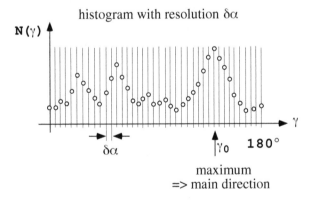

Figure 4.13 Angle histogram of a room

Φ gives the direction of the walls with respect to the vehicle axis. In order to emphasize walls far away rather than those close by, α_i is counted with the weight $w_i = (r_i + r_{i+k})/2$:

$$n_j = \sum_{j=j_1}^{j_2} w_i \quad \text{with} \quad j\delta\alpha \leq \alpha_i < (j+1)\delta\alpha \tag{4.39}$$

The next step is an alignment of the vehicle axis to the structure of the room: Take the largest of the peak angles and rotate the picture of the room by this angle. Then the picture is aligned to one of the main directions of the room and the peak angle is the direction of the vehicle with respect to the room. For indoor systems this is a much better description than the alignment with respect to true north. The point cloud is now described in a coordinate system centered around the vehicle but aligned to the room.

Box feature Two new histograms projecting the points to the x-axis and the y-axis are set up. The distances of the peaks directly give the dimensions of the rectangular room. This is an invariant of the room and a landmark in itself. The user is then asked to give a "name" to this box and later on the position of the vehicle can be described in a natural way: as being in room "name". As the dimensions of the room are larger than the dimensions of the vehicle, it is much simpler to find the room again than the exact position of the former. Figure 4.14 shows an example.

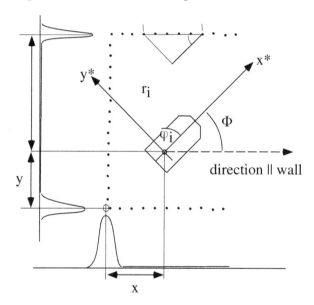

Figure 4.14 Determination of the box feature

Edge extraction It it also possible to extract edges from the point cloud: the angles are calculated as described before but form a data vector instead of a histogram: ($\rightarrow \phi_i$) and describe the distances as a the distance vector ($\rightarrow r$). If $(r_{i-1} - r_i) \gg (r_i - r_{i+1})$ or $(r_{i-1} - r_i) \ll (r_i - r_{i+1})$, a discontinuity in the environment is detected. This might be an edge or two objects obscuring each other. Edges can also indicate themselves in sudden changes in the measured angles. If $\phi_i >> \phi_{i+k}$ or vice versa, an edge in the environment has been detected. By (r_i,ϕ_i) the coordinates of the edge are given and may be used as landmarks.

4.5 Landmarks

Landmarks are easily detectable features in a scene which make it possible to find the position and orientation of a vehicle.

4.5.1 Natural landmarks

Natural landmarks could be anything from the box feature in an indoor environment, edges of stationary objects to markers found in the environment or readily recognizable objects.

4.5.2 Artificial landmarks

While boxes and edges are natural landmarks, the environment might also be equipped with artificial landmarks. Examples for the latter are lighthouses on a coast or easily recognizable markers brought into the environment to ease finding positions, like barcode strips. They directly tell their position but are rather disturbing in a living environment. Better landmarks making recognition easier for humans are numbers on a door or name plates. In our own home we would dislike the idea of mounting artificial markers. Thus other hints are needed to tell us our position with respect to the environment. This leads into the vast area of object recognition, handled later on in this book.

4.5.3 Triangulation using landmarks

To find out position and orientation in closed rooms where GPS cannot be used, other landmarks are needed. Those could be either natural or artificial landmarks both active or passive. Sensors on board the vehicle must be able to detect these marks. With active landmarks the task of finding position and orientation is rather simple: they emit some sort of energy like light preferably in the near infrared, so they do not disturb human beings in a room. The light is coded with a characteristic number unique to this specific landmark. This light is then detected by a rotating sensor on board the vehicle. It provides the angle between a symmetry axis of the vehicle and the source of light as well as the number of the landmark as output signal. The same principle was used for lighthouses in the 19th and 20th century. From three angles measured, the position and orientation of the vehicle with respect to the room can be deduced. Let P_1, P_2 and P_3 be three landmarks with coordinates (x_1,y_1), (x_2,y_2) and (x_3,y_3) respectively, not arranged in a line. The measured angles are α, β and γ. Let P be the position of the vehicle itself with coordinates (x,y). Figure 4.15 depicts the situation described. $P(x,y)$ is located on a circle through P_1 and P_2. The difference angles $\phi_{12} = \beta - \alpha$ and $\phi_{23} = \gamma - \beta$ are invariants under the rotation of the vehicle. The radius of the circle is R_1.

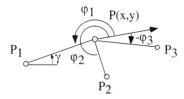

Figure 4.15   Position from angles to three landmarks

Let $a*$ be the distance between P_1 and P_2, then

$$R_1 = \frac{a*}{2\sin \phi_{12}} \tag{4.40}$$

$$a* = \sqrt{(x_2 - x_1)^2 + (y_2 - y_1)^2} \tag{4.41}$$

The midpoint of $a*$ is PM with its coordinates

$$x_m = \frac{x_2 - x_1}{2} \tag{4.42}$$

$$y_m = \frac{y_2 - y_1}{2} \tag{4.43}$$

With

$$z = R_1 \cos \phi_{12} \tag{4.44}$$

$$\tan \alpha_1 = \frac{y_2 - y_1}{x_2 - x_1} \tag{4.45}$$

the coordinates of the midpoint M_1 of the circle are

$$x_{M_1} = x_m + z \cos \alpha_1 \tag{4.46}$$

$$y_{M_1} = y_m + z \cos \alpha_1 \tag{4.47}$$

The position of P looked for then has coordinates implicitly given by the circle equation

$$R_1^2 = (x - x_{M_1})^2 + (y - y_{M_1})^2 \qquad (4.48)$$

Accordingly, for the circle through P, P_2 and $P3$ including an angle ϕ_{13} as shown in figure 4.16 there is

$$b* = \frac{x_3 - x_2}{\sin \alpha_2}, \qquad (4.49)$$

$$R_2 = \frac{b*}{\sin \phi_{23}}, \qquad (4.50)$$

$$\tan \alpha_2 = \frac{y_3 - y_2}{x_3 - x_2} \qquad (4.51)$$

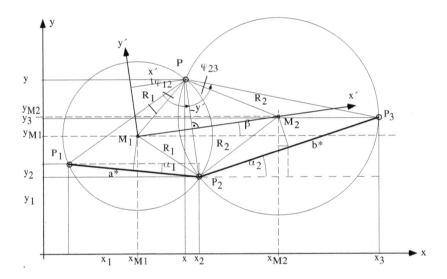

Figure 4.16 Triangulation using landmarks

The coordinates of the midpoint of the circle through P, P_2 and $P3$ are

$$x_{M_2} = \frac{x_3 - x_2}{2} - \frac{b*}{2 \cot \phi_{23} \sin \alpha_2} \qquad (4.52)$$

$$y_{M_2} = \frac{y_3 - y_2}{2} + \frac{b*}{2 \cot \phi_{23} \cos \alpha_2} \qquad (4.53)$$

The calculation of the coordinates of the point P is first done in a coordinate system (x',y') running through the points M_1 and M_2 and with M_1 as center point as shown in figure 4.17. Let L be the distance between M_1 and M_2 then

$$L = \sqrt{(x_M2 - x_M1)^2 + (y_M2 - y_M1)^2} \tag{4.54}$$

The point P lies on a circle with radius R_1 around M_1:

$$((x')^2 + (y')^2) = (R_1)^2 \tag{4.55}$$

and also on a circle with radius R_2 around M_2:

$$\left((L - x')^2 + (y')^2\right) = (R_2)^2 \tag{4.56}$$

The coordinates of point P are thus

$$x' = \frac{(R_1)^2 - (R_2)^2 + L^2}{2L} \tag{4.57}$$

$$y' = \sqrt{(R_1)^2 - (x')^2} \tag{4.58}$$

The last step in this calculation is the coordinate transformation into the world coordinate system (x,y) according to figure 4.17. Errors in the angles measured show up in circles of different radii and give rise to an error quadrangle as shown in figure 4.18

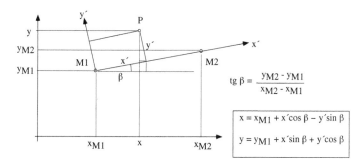

Figure 4.17 Coordinate transformation

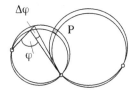

Figure 4.18 Position error by error in angles

4.5.4   Measuring distances to artificial landmarks

If the distances to artificial landmarks can be obtained, the position of the vehicle is much simpler to deduce. According to figure 4.19, distances R_1 to landmark P_1 and R_2 to landmark P_2 are measured from the current position P. P is thus the cutting point of a circle with radius R_1 around P_1 and R_2 around P_2. The calculation gets simple in a coordinate system x' and y' through P_1 and P_2 according to figure 4.20.

$$x' = \left(R_2^2 - R_1^2 + L^2 \right) / 2L \tag{4.59}$$

$$y' = \sqrt{\left(R_1^2 - x'^2 \right)} \tag{4.60}$$

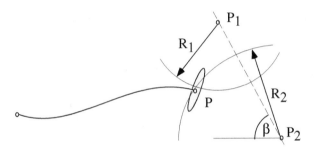

Figure 4.19   Measuring distance to landmarks

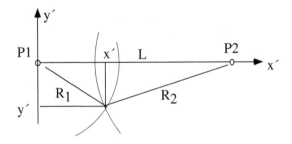

Figure 4.20   Position from measured distances

The transformation into the world coordinate system follows figure 4.17. An error in position will stem from errors in distance measurements forming an error square of $\Delta R_1 \times \Delta R_2$ around position (x,y).

4.5.5   Artificial active landmarks

Artificial active landmarks send out signals so that a vehicle may find its position and orientation from analyzing these signals.

- Lighthouses on a coast are examples: they send out their signature in regular intervals

- Two parallel lasers rotating at a constant angular velocity allow the calculation of the distance to that landmark. Figure 4.21 shows an example [PKEv00]

- A rotating sheet of light successively hitting three sensors on board a vehicle allows the measurement of position and orientation simultaneously. Figure 4.22 and figure 4.23 show the principle. From the known rotation rate of the landmark beam, the starting time when the beam angle is zero with respect to the x-axis and the triangle side length the distance r between the coordinate center (0,0) of the light house and the kinematic center of the vehicle, the angle φ between the x-axis and the kinematic center and the orientation of the vehicle $\psi = \varphi - \alpha$ may be deduced. The calculation is left to the reader as an exercise.

- Two parallel lasers and one single laser rotating at a constant angular velocity allow the calculation of the position of the robot.

- Satellites sending out their position at regular intervals allow the worldwide determination of positions (see also section 4.6).

To determine not only the distance between the rotating laser beams (artificial active landmark) and the robot but also the position of the robot due to the coordinate system of the landmark, it is necessary to extend the system by an additional laser. This laser is counterrotating to the twin laser as shown in figure 4.24. On the vehicle a plastic fiber antenna is fixed, receiving the optical impulses of the laser points. Based on the time shift between the laser impulses, the rotation frequency of the disks f and the displacement of the twin laser a the distance d and the orientation w of the robot to the landmark coordinate system is calculated (see equation 4.62). The parameters used for the calculation are presented in figure 4.25.

$$\text{distance:} \quad d = \frac{a}{2 \sin\left(\frac{T_{Zw}}{T}\right) \cdot \pi} \tag{4.61}$$

$$\text{angle:} \quad w = 180° \cdot \frac{T_{Zw-E} + \frac{T_{Zw}}{2}}{T} \tag{4.62}$$

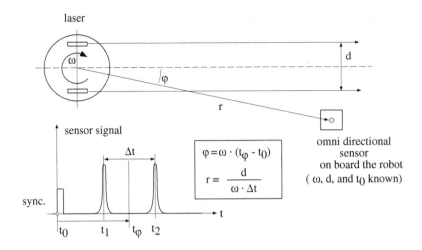

Figure 4.21 Twin-laser system

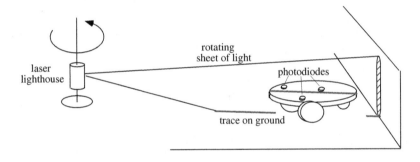

Figure 4.22 Rotating sheet of light

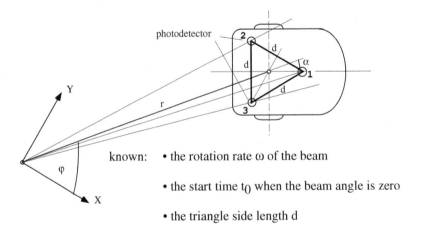

Figure 4.23 Sheet of light hits three photosensors

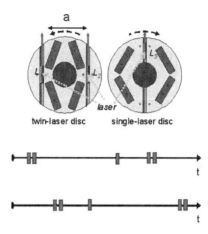

Figure 4.24 Measurement concept of an active landmark with 3 lasers. The position of the lasers on the rotating disks is shown as well as the diagram of the received light impulses over time.

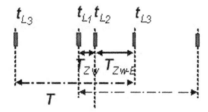

Figure 4.25 Parameter for the calculation of position of the vehicle. T_{Zw} is the time shift between the twin lasers, T_{Zw-E} determines the time shift between twin and single laser, and T duration for one rotation of the disks.

It is obvious that there exist positions in which the measurements are ambiguous. In figure 4.26 a typical situation is shown in which it is not possible to determine the correspondence between the laser and detected laser impulse. This means that there are situations in which the position cannot be calculated. For practical use of the measurement system, one can calculate the ambiguous areas. If the robot is located in these areas additional information like odometry or pose of the robot during the recent measurements have to be used to solve the ambiguity.

In [Hac06] such a localization system for a small forklift robot is described. This laser beacon consists of two disks with diameters of 160 mm, a distance between the disks of 6 mm and the distance a between the twin lasers of 100 mm. The actuation system for the disks is realized with DC motors. A rotational frequency of the disks of 5 Hz is used for the experiment presented below. The laser diode generates a laser point of 5 mm diameter. The divergence of the laser beam is 0.5 mrad. Because the lasers have to be eye safe, the power consumption of the laser diodes are less than 2 mW. In figure 4.27, the laser beacon and the forklift robot are shown.

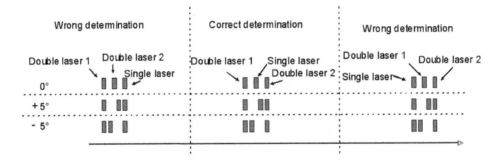

Figure 4.26 Ambiguity of the signals. The angles in the diagram are corresponding to the situation that the twin and single laser are oriented in the same direction.

Figure 4.27 Laser beacon and the placement of the antenna on the forklift robot

Experiments to determine a position expressed in polar coordinates show that the angle precision is less than ±0.5°. Based on the measurement principle the derivation of the radius is ± 0.5 cm (below 1 m) and ± 9 cm (at 17 m). In figure 4.28 several experiments for different distances are presented with the deviation. The precision of the system can be improved by the optimization of the actuation system (constant rotational velocity of the disks) and fixing of the lasers (the twin lasers have to be parallel).

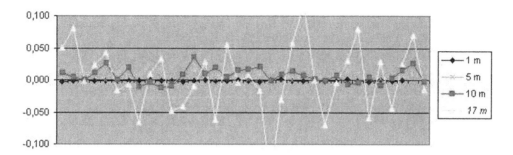

Figure 4.28 Distance and deviation (in m) of several measurements up to 17 m is shown

4.6 Global positioning system (GPS)

Born from military needs during the 1970s, the US Department of Defense installed a system of 24 satellites in 6 orbit levels at a height of 20.051 km. Each satellite covers 1/6 of the earth surface and they are arranged at such orbits that the 24 satellites cover each point on earth fourfold, i. e. at least four satellites can be seen at an arbitrary moment from every point on earth.

The satellites are equipped with an atomic clock and synchronized with each other. They are sending out signals containing the number of the satellite, its position at the moment of the timing signal, and a timing signal itself. The signals are picked up by earthbound receivers. Using the signals from four satellites, a receiver can find out its position on earth measuring the time differences between sender signals, given their positions known at the moment of a timing signal. Intended in the beginning for military purposes only, a crude form of the system was made available to the public after the first Gulf war and became a tremendous success. It operates worldwide with an accuracy of approximately 10 m. It is used in navigation systems in cars, ships, and on board of agricultural machines. Even hikers may use it to find their way through the wilderness or in the maze of a medieval town.

Evaluating the time differences between the position signals from four satellites, three sources of position information may be gathered: the longitude, the latitude and the height above ground. So at any time at least four satellites should be visible. Using the measured time differences, the calculation for the Standard Positioning Service (SPS) can be performed solving the following equation system with four unknowns for the receiver coordinates with satellite's coordinates (SV), receiver's coordinates (R), and distance receiver/satellite (D):

$$\sqrt{(SV_{x0} - R_x)^2 + (SV_{y0} - R_y)^2 + (SV_{z0} - R_z)^2} = D_0 + \Delta D$$

$$\sqrt{(SV_{x1} - R_x)^2 + (SV_{y1} - R_y)^2 + (SV_{z1} - R_z)^2} = D_1 + \Delta D$$

$$\sqrt{(SV_{x2} - R_x)^2 + (SV_{y2} - R_y)^2 + (SV_{z2} - R_z)^2} = D_2 + \Delta D$$

$$\sqrt{(SV_{x3} - R_x)^2 + (SV_{y3} - R_y)^2 + (SV_{z3} - R_z)^2} = D_3 + \Delta D$$

Deviations are unavoidable and are dependent on reference satellite alignment. The best results are obtained at a 90° angle between the receiver and the satellite, the worst in a situation where all available satellites are located in close proximity to each other or are in a collinear alignment. Possible position errors could emerge from satellite position (3 m), ionosphere refraction (5 m), troposphere refraction (2 m), multiple reflection (5 m), and a "selective availability" (30 m). Their microwave signals may be obscured by trees or high rising buildings and vanish inside buildings or in tunnels.

There are other systems planned or being installed: The Russian GLO-NASS, to become operative in 2009 and the European Galileo-system, planned to start operation in 2013, the latter with 64 satellites and a position error of less than 1 m.

To enhance distance resolution, two GPS receivers can be used. One is positioned at a known fixed location and the other one at an unknown position nearby at a distance of less than a few kilometers apart. The measured signal at the fixed location and the signal measured at the unknown position are compared. The latter may be calculated then to fractions of a centimeter resulting in a so called differential GPS. Measuring a new position every 100 ms, the orientation of a vehicle may also be calculated. With the receiver and the calculating electronics shrunken to the size of a cigarette box and equipped with a digital map of the region of interest, the GPS has solved the problem of localization outdoors in open terrain. The landmarks used are the known positions of satellites transferred to a navigation system via precise timing signals.

4.7 Kalman filter

Kalman filters are optimal estimators for unobservable system states if some preconditions are met. These conditions are

- the system and measurement models have to be linear,

- the additive noise of system and measurement models have to be "white" and Gaussian distributed.

If and only if these conditions are met, the filter guarantees to generate the optimal estimate for the modelled system. In fact it turns out that system transitions and sensor errors are very well described using linear models and Gaussian noise models. This enables the user to apply a Kalman filter even if the required properties are approximated. The Kalman filter has shown to be stable and correct in a broad range of applications. One of these applications is the localization of mobile robots.

4.7.1 General idea

More often than not a system is given with observable outputs $\vec{y}(t_k)$ at time step t_k but an unobservable internal state $\vec{x}(t_k)$. The idea of a Kalman filter is to build a model of this system in which the internal state is observable and to correct this state comparing the output of the real system and the output of the model. The system to be modeled is described by

$$\vec{y}(t_k) = \begin{pmatrix} y_1(t_k) \\ \vdots \\ y_n(t_k) \end{pmatrix} \qquad \vec{x}(t_k) = \begin{pmatrix} x_1(t_k) \\ \vdots \\ x_m(t_k) \end{pmatrix} \qquad (4.63)$$

and an input vector $\vec{u}(t_k)$. Let the operation of the system be as sketched in figure 4.29: With an initial state $\vec{x}(t_0)$ and matrices $A(t_k)$ and $B(t_k)$ and $H(t_k)$ the next internal state is $\vec{x}(t_{k+1})$.

$$\vec{x}(t_{k+1}) = A(t_k)\vec{x}(t_k) + B(t_k)\vec{u}(t_k) + \vec{v}(t_k) \quad . \qquad (4.64)$$
$$\vec{y}(t_k) = H(t_k)\vec{x}(t_k) + \vec{w}(t_k) \qquad (4.65)$$

There is inevitable noise in the system: $\vec{v}(t_k)$ is the system noise and $\vec{w}(t_k)$ the measurement noise. To get the still unknown internal state $\vec{x}(t_k)$, a noise free model is run in parallel to the real system: with matrices A,B and H and an initial value $\vec{x}^*(t_0) = 0$.

The output of the model is

$$\vec{y}^*(t_k) = H(t_k)\vec{x}^*(t_k) \qquad (4.66)$$

Let A,B and H be treated as time invariant: The model internal state $\vec{x}^*(t_k)$ is enhanced using the measured output $\vec{y}(t_k)$ and a Kalman amplification $K(t_k)$:

$$\vec{x}^{**}(t_k) = \vec{x}^*(t_k) + K(t_k)(\vec{y}(t_k) - \vec{y}^*(t_k)) \qquad (4.67)$$

Kalman filter

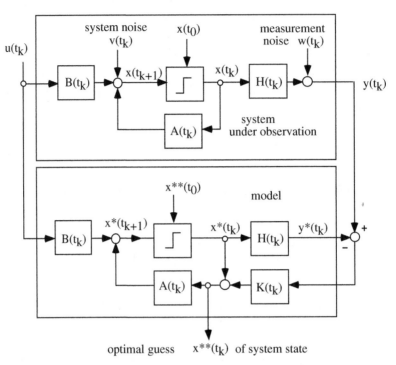

Figure 4.29 Overview of the Kalman filter: The unobservable system is modelled using the system model A, the input model B and the measurement model H. The difference between estimated and measurement output influences the Kalman Gain K which realizes an optimal weighting factor between observation and estimation.

Then the next internal value of the model is calculated

$$\vec{x}^*(t_{k+1}) = A(t_k)\vec{x}^{**}(t_k) + B(t_k)\vec{u}(t_k) \tag{4.68}$$

The Kalman amplification is calculated so that the sum of the variances of the error gets minimal: Let P be the covariance matrix of the guessing error $E[(\vec{x} - \vec{x}^{**})(\vec{x} - \vec{x}^{**})^T]$. Then the trace of that matrix shall be minimal:

$$\text{trace}(P) = \sum_{j=1}^{n} \sigma_i^2 = \text{minimal} \tag{4.69}$$

4.7.2 Guessing error

If $\text{trace}(P) = f(K)$, then $\partial(\text{trace}(K))/\partial K = 0$ minimizes the sum of the guessing errors and gives an equation for K. The guessing error is $(\vec{x} - \vec{x}^{**})$.

Let each of its components $(x_i - x_i^{**})$ be normally distributed. Then the probability (see also figure 4.30) to find in the component i the error $(x_i - x_i^{**})$ is

$$p(x_i - x_i^{**}) = \frac{1}{\sigma_i \sqrt{\pi}} \exp^{[(x_i - x_i^{**})/\sigma_i]^2} \qquad (4.70)$$

with

$$\int_{-\infty}^{+\infty} p(x_i - x_i^{**})d(x_i - x_i^{**}) = 1 \qquad (4.71)$$

σ_i^2 is the variance of the guessing error of component i; it is the awaited value of $(x_i - x_i^{**})^2$ or the mean value of M measurements for $M >> 1$.

The theory of a Kalman filter shows how to calculate the Kalman amplification from the given variances of the measured output values. This gives a rather good enhancement of the calculation of the vehicle pose and shrinks the remaining error considerably.

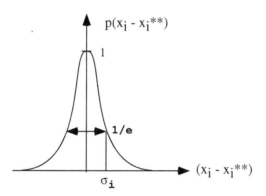

Figure 4.30 Normal distribution of an error

4.7.3   Example application

Let us assume that a mobile robot is driving along a very long and straight corridor. The one dimensional position of the robot can be measured with a set of ceiling-mounted cameras. Unfortunately there is a gap in the camera views. In these gaps the position of the robot is not observable. Assume further that position and speed of the robot in that corridor should be estimated using a linear discrete Kalman filter.

The state vector is defined to be $x(t_k) = [p,s]^T$ with position p and speed s. The measurement vector consists only of the position and is defined as $y(t_k) = [p]$. The system model uses the basic motion equation $p(t_{k+1}) = p(t_k) + \Delta t \cdot s(t_k)$. The motion equation results in the system model

$$A = \begin{bmatrix} 1 & \Delta t \\ 0 & 1 \end{bmatrix} \quad \Rightarrow$$

$$\vec{x}(t_k) = A \cdot \vec{x}(t_{k-1}) \tag{4.72}$$

$$\begin{pmatrix} p(t_k) \\ s(t_k) \end{pmatrix} = \begin{pmatrix} p(t_{k-1}) + \Delta t \cdot s(t_{k-1}) \\ s(t_{k-1}) \end{pmatrix} \tag{4.73}$$

The measurement model simplifies to $H = [10]$ since the position is measured directly and the speed is not measured at all. The input model is set to $B = [0]$ since no inputs are used.

The selection of the noise vectors $\vec{v}(t_k)$ and $\vec{w}(t_k)$ is, at least to some extent, based on experiments and knowledge. The system noise in this example is chosen to be very small since it is assumed that the velocity of the robot is constant between two time steps. In comparison to the system noise, the measurement noise is variable since it heavily depends on the observation. In this example it is chosen to increase dramatically if the robot is not visible.

$$\vec{v}(t_k) = \begin{pmatrix} 0.00001 \\ 0.00001 \end{pmatrix} \tag{4.74}$$

$$\vec{w}(t_k)^{\text{visible}} = (1) \tag{4.75}$$

$$\vec{w}(t_k)^{\text{not visible}} = (10) \tag{4.76}$$

Now assume the following experiment: A mobile robot is driving down a corridor of 100 m length. The speed of the robot is constant $0.5\,\frac{m}{s}$. The corridor is equipped with a set of cameras for position measurements with an unobservable gap between 25 m and 50 m. The cameras measure the position once a second which results in 200 measurements during the whole experiment. Whenever the robot is not observable, the last known position is returned.

Figure 4.31 plots the output of the Kalman filter during the experiment. It can be seen that the state of the filter converges to the true position and speed about measurement 20. Between measurement 50 and 100, the measurement gap occurs. It is obvious that both position and speed are divergent and the position variance is increasing. After time step 100 the filter is again converging to the correct values.

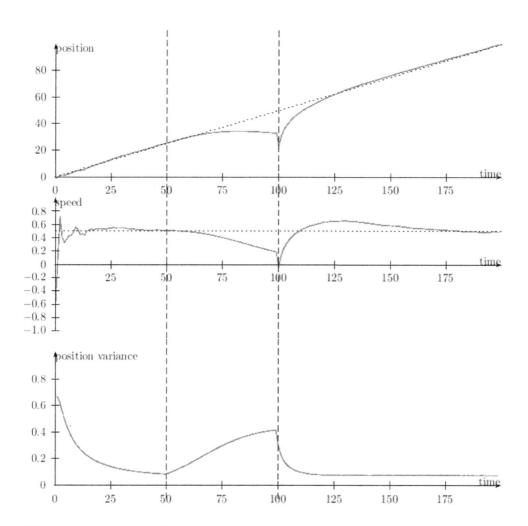

Figure 4.31 Output of the Kalman filter during the localization of a mobile robot in a corridor. The diagrams show the estimation of the state (position and speed) and the variance of the position. It can be seen that between time step 50 and 100 the variance is increasing and position and speed are not converging to the dotted lines. During that period, position measurements are not available and the estimation error is dramatically increasing.

5 Mapping

Whenever a mobile robot is required to navigate beyond its sensory horizon, it must either rely on potentially ineffective or misleading local search strategies (such as the 'bug algorithms' [Lum87]) or use some kind of world model to store cues for navigation. Such a world model is generally called a 'map' and can either be provided a priori or built online using a mapping algorithm. The mapping approaches can be separated into world-centric or robot-centric. World-centric systems represent the pose of all objects including the robot of the environment according to a fixed coordinate frame. In indoor scenarios, a corner of a room or a fixed position in the entrance area of an apartment is often used. To specify positions in the operational environment of the robot in outdoor applications, global coordinate systems like the latitude, longitude, and height system, the Earth Centered, Earth Fixed Cartesian coordinate system, the World Geographic Reference System or WGS 84 (GPS) are often used. World-centric mapping is mainly employed for tasks like navigation or path planning while robot-centric approaches are used for piloting tasks such as collision avoidance. Using matrix-based coordinate transformations, it is possible to convert between these different reference frames.

The main problem of generating maps is the inaccuracy of the sensor systems being used for solving the localization problem and measuring objects in the environment of the robots. Therefore, it is very difficult to build global maps based on local ones, to update existing maps or to correspond the objects stored in the map with those measured in the environment of the robot.

Concerning map types, there are numerous ways to model the environment of a vehicle. The data structures found in literature can be broadly divided into the four classes of *metrical, grid, topological* and *hybrid* maps (see figure 5.1).

Purely metrical maps are probably the most common type. Systems relying on these maps are characterized by using one global, metrically consistent frame of reference. The accuracy of the stored map is approximately equal to the quality of available sensor data. Also, all metrical locations are equally important. Metrical maps comprise geometric features and their spatial locations. The features actually used can range from very basic (such as 3D points calculated from range measurements) over geometrically more expressive line or box features up to semantically very distinct landmarks, which can be uniquely identified from a large body of sensor data.

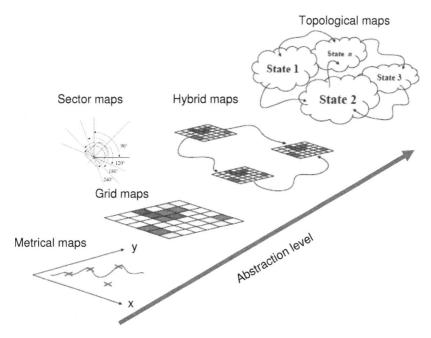

Figure 5.1 Abstraction level of the different mapping approaches

Grid maps, which are very often used for mobile robots, are popular due to their simplicity and intuitive representation. A grid map divides space into adjacent portions of equal metrical sizes. For a two-dimensional map, this results in a square grid, while the three-dimensional grid map resembles a Rubic's cube. Both dimensionalities have been used [Elf89, Mor96], but three-dimensional grid maps are rare due to their excessive storage requirements. Two major variants of grid maps are occupancy grids and elevation maps.

Unfortunately, the highly detailed metrical world representation requires a lot of memory and leads to algorithms with high computational demands. These properties limit the scalability of both grid and metrical maps [Bro87].

Motivated by these drawbacks, researchers aiming at large scale navigation early on began looking at the *topological* world model, which represents the environment in a more compact, qualitative fashion. Topological approaches focus on representing navigation-relevant places and their connections on an abstract level rather than the exact metrical layout of the surroundings. Thus, imprecise localization is less of a problem for topological approaches and algorithms can run faster because they have to cope with much less data. Topological maps commonly use graphs as their underlying data structure. Graph nodes identify locations of interest and their characteristic features, while knowledge about travel between nodes is encoded in the connecting graph edges.

More recently, many researchers have proposed to attack the problems of autonomous mapping using combinations of the metrical and topological methodologies. Generally, these *hybrid* approaches are designed to combine the benefits of both representational forms, ideally allowing localization and map building with the high precision of metrical maps while retaining the computational tractability and compactness of topological data structures.

In the following, several successful examples of mapping techniques using different map types are presented.

5.1   Metrical maps

In metrical maps, the environment is described with the help of geometrical features. These geometrical features can be 2D or 3D points, lines, polygons or 2D areas like rectangles. As an example for indoor scenarios, lines could describe the walls of rooms. In outdoor scenarios, lines could denote streets or highways.

Metrical mapping offers several advantages. Geometrical features can be maintained over time even if their positions change. Thus it is obvious that this type of map is also suited for dynamic environments. Another advantage is that metrical feature maps offer a more compact description of the surroundings than grid maps. Hence they are superior especially in a scenario with a rather structured environment.

In the following the line-based and plane based metrical maps, which can both be generated by a robot based on distance measurements, are introduced.

5.1.1   Line based metrical maps

Lines extracted from distance measuring sensors form borders between regions that can be divided into either free space, obstacles or unknown regions. Obstacles may be represented by clusters or border lines. In the following, a method is presented to extract both from (laser) distance measurements.

Line segmentation The idea of the line segmentation technique is to cluster a radar scan into groups of distance points that lie close together. Let us take a radar scan $\{r_i, \varphi_i\}$ with $i = 1, \ldots, n$ measured from position $Q = (x_0, y_0, \psi)$. Let $P_i = (r_i, \varphi_i)$. As results, corners C_j, segments S_k and auxiliary clusters H_m will be formed.

The procedure is shown in algorithm 5.1.

Algorithm 5.1 Line segmentation

Initialization: $i := 1; j := 1; k := 1; m := 1; H_1 := (P_1)$
while $i < n - 1$ **do**
  if $|P_{i+1} - P_i| < d$ **then**
    $H_m := H_m \cup P_{i+1}; i := i + 1;$
  else if $|P_{i+2} - P_i| < d$ **then**
    $H_m := H_m \cup P_{i+2}; C_j := (P_{i+1}); j := j + 1; i := i + 1;$
  else if $|P_{i+2} - P_{i+1}| < d$ **then**
    $m := m + 1; H_m := (P_{i+1}, P_{i+2}); i := i + 2;$
  else
    $C_j := (P_{i+1}); j := j + 1; m := m + 1; H_m := (P_{i+2});$
  end if
end while
while $m > 0$ **do**
  if number in $(H_m) \leq c$ **then**
    $C_j := H_m; j := j + 1; m := m - 1;$
    (* c is the maximal number in a cluster*)
  else
    $S_k := H_m; j := j + 1; m := m - 1;$
  end if
end while

Iterative end point fit [DH73]
Take a segment S_k with points $P_q, \ldots, P_r$ according to figure 5.2 and form the distances $h_{r-1}, \ldots, h_{q-1}$. They are calculated following figure 5.3 for two points P_q and P_r. Let ψ be the angle of the line connecting both points with respect to the common coordinate system.

$$\tan \psi = \frac{y_q - y_r}{x_q - x_r} \tag{5.1}$$

$$h_i = ((y_i - y_q) - (x_i - x_q) \tan \psi) \cos \psi \tag{5.2}$$

$$\Rightarrow \boxed{h_i = (y - i - y_q) \cos \psi - (x_i - x_q) \sin \psi} \tag{5.3}$$

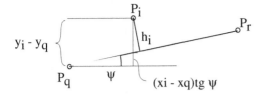

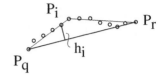

Figure 5.2 Distance to line   **Figure 5.3**   Iterative endpoint fit

Take the line between P_q and P_r. If $h_j = \max(h_i) > \varepsilon$ for $r < i < q$; (*ε is the allowed fuzziness in deviations from a line*) then form two lines (P_r, P_j) and (P_j, P_q) and repeat the calculation, else the line (P_r, P_q) is established. The effort is equal to the number of lines to be inserted between P_q and P_r.

Regression line   Find a line through N points $P_q, \ldots, P_r$ so that the sum of the squared distances to that line gets minimal. There are $h_i = (y_i - y_q) \cos \psi - (x_i - x_q) \sin \psi$ the distances to that line. Set for a moment $P_q = (0, b)$, then the distances become $h_i = (y_i - b) \cos \psi - x_i \sin \psi$.

The sum of the squares is

$$\sum_i h_i^2 = \sum_i (y_i - b)^2 \cos^2 \gamma - 2(y_i - b) x_i \sin \gamma \cos \gamma + x_i^2 \sin^2 \gamma \qquad (5.4)$$

This is to be minimized with respect to b:

$$\frac{\partial(\sum (\ldots))}{\partial b} \stackrel{!}{=} 0$$

$$\sum_i -2(y_i - b) \cos^2 \gamma + 2x_i \sin \gamma \cos \gamma \stackrel{!}{=} 0 \qquad (5.5)$$

$$\sum_i -(y_i - b) \cos \gamma + x_i \sin \gamma = 0 \qquad (5.6)$$

$$\sum_i b - y_i + x_i \tan \gamma = 0 \qquad (5.7)$$

$$N \cdot b = \sum_i y_i - \tan \gamma \sum_i x_i \qquad (5.8)$$

$$y_s = \tan \gamma x_s + b \qquad (5.9)$$

This is a line through the center of gravity of the points. The next step now is to minimize the squares of distances: The squares of the distances are

$$\sum_i h_i^2 = \sum_i (y_i - b)^2 \cos^2 \gamma - 2(y_i - b) x_i \sin \gamma \cos \gamma + x_i^2 \sin^2 \gamma \qquad (5.10)$$

The angle γ is to be minimized:

$$\frac{\partial(\sum(\ldots))}{\partial\gamma} \overset{!}{=} 0$$

$$\sum_i^N -2(y_i - b)\cos\gamma\sin\gamma - 2(y_i - y_s)(x_i - x_s)(-\sin^2\gamma + \cos^2\gamma)$$

$$+ \sum_i^N 2(x_i - x_s)^2\sin\gamma\cos\gamma \overset{!}{=} 0 \quad (5.11)$$

$$\sum_i^N (y_i - y_s)^2\tan\gamma + (x_i - x_s)(y_i - y_s)(1 - \tan^2\gamma) + (x_i - x_s)^2\tan\gamma = 0$$

$$(5.12)$$

$$\sum_i^N (y_i - y_s)(x_i - x_s)\tan^2\gamma + \sum_i^N (-(y_i - y_s)^2 + (x_i - x_s)^2)\tan\gamma$$

$$- \sum_i^N (y_i - y_s)(x_i - x_s) = 0 \quad (5.13)$$

The last equation is a quadratic equation in $z = \tan\gamma$:

$$a \cdot z^2 + b \cdot z - a = 0 \quad (5.14)$$

$$a = \sum_i^N (y_i - y_s)(x_i - x_s) \qquad b = \sum_i^N (y_i - y_s)(x_i - x_s) \quad (5.15)$$

$$\boxed{z_{1,2} = -\frac{b}{2a} \pm \sqrt{1 + b^2/4a^2}} \quad (5.16)$$

Shift of line end points   Given a line through the center of gravity for a couple of points, find proper endpoints Q and R in the vicinity of P_q and P_r. Figure 5.4 shows the situation.
Let $\psi \approx \gamma$. Then

$$h_s = (y_s - y_q)\cos\psi - (x_i - x_s)\sin\psi \quad (5.17)$$

$$\sum h_i - h_s = \sum (y_i - y_s)\cos\psi - (x_i - x_s)\sin\psi \quad (5.18)$$

$$\sum h_i - h_s = \cos\psi(\sum y_i - N \cdot y_s) - \sin\psi(\sum x_i - N \cdot x_s) = 0 \quad (5.19)$$

This means a shift of the line end points by h_s:

$$P_q \longrightarrow Q = (x_q + \Delta x, y_q + \Delta y) \qquad \Delta x = h_s\sin\psi \quad (5.20)$$

$$P_r \longrightarrow R = (x_r + \Delta x, y_r + \Delta y) \qquad \Delta y = h_s\cos\psi \quad (5.21)$$

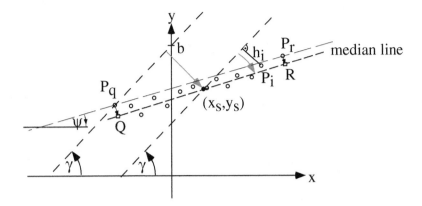

Figure 5.4 Fitting the regression line

Fusion of lines   Let the vehicle take radar pictures from different positions P_0 and P_1. Let (P_i, P_k) be a line extracted from radar scan R_1 at P_0 and (P_i', P_k') extracted from radar scan R_2 at P_1. Let the angles be ψ and $\psi' = \psi + \varepsilon$. Figure 5.5 shows two cases of fusions:

- Let $h_i' < \delta$ for P_i' with respect to the line (P_i, P_k) and P_i' between P_i and P_k and moreover $|P_i', P_k| > |P_i', P_k|$ then both lines are condensed into one line (P_i, P_k').

- Let $|P_k, P_i'| < d$ then both lines are fused into one line (P_i, P_k'), too.

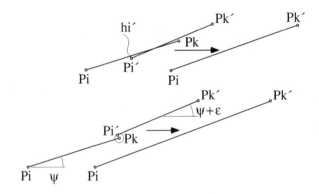

Figure 5.5 Fusion of lines

Obstacles vs. free space Introducing a direction in a line allows to differentiate between free space and possible obstacles for a vehicle, as shown in figure 5.6: to the right of a line from P_1 to P_2 there is a free space open, otherwise it could not have been constructed. To the left is unknown territory, possibly the line is the border of an obstacle. A room may be described by a polygon with lines oriented clockwise; they might also describe isolated obstacles if the lines are oriented counter clockwise.

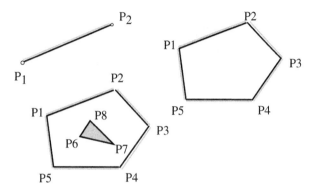

Figure 5.6 Defining free space

Given a radar scan with some lines extracted, a certain amount of free space is established. Driving around, this free space can be enlarged as shown in figure 5.7. In general some parts of the environment remain unknown.

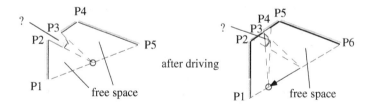

Figure 5.7 Enlarging free space by driving around

5.1.2 Plane based metrical maps

Planes are often used as features to define reliable landmarks such as floor, ceiling, walls, doors and big objects of furniture like desks and cupboards. The main problem of this environment representation is the extraction of planes. A typical plane extraction algorithm [WGS03] is based on a 3D point cloud. The corresponding data acquisition is based on 3D sensor systems as rotating 2D laser scanners (see figure 5.8) or time-of-flight cameras. Figure 5.9 shows a typical 3D point cloud of a real cluttered indoor scene.

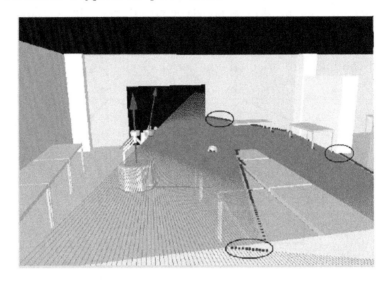

Figure 5.8 3D data acquisition in a virtual indoor scene. The laser beams are visualized as rays and the measurement points as black dots.

Figure 5.9 3D point cloud of a typical indoor scene

The goal is to approximate the input data by a set of planar patches so that each set of points is optimally represented in a least square sense by its plane. Features represented by a large amount of samples (e. g. corridor walls) are reduced to one large planar patch. Figure 5.10 shows a result of this process applied to the 3D point cloud shown in figure 5.9.

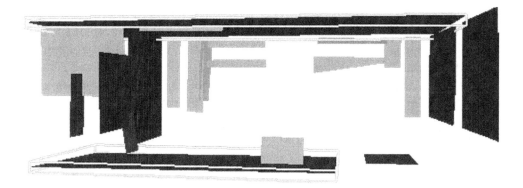

Figure 5.10 Planes extracted from the 3D point cloud of figure 5.9. The ground and ceiling plane are marked by a bounding box.

The following gives a summary of the different steps of the extraction method. The complete strategy is described in more detail in [Ast07]. Figure 5.11 shows a flow chart of the whole procedure and algorithm 5.2 sketches the different processing steps.

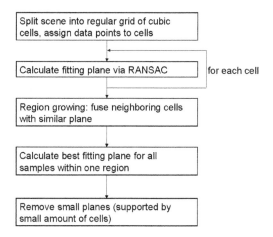

Figure 5.11 Flow chart of RANSAC based plane extraction

In step 1 and 2 the whole point cloud is split into a regular grid of cubic cells (see figure 5.12) in order to perform plane fitting locally. In this way smaller features as chairbacks and screens can be extracted. Big planes from floor, walls and ceiling are detected via region growing in step 4. The RANSAC algorithm in step 3 repeatedly calculates planes approximating the local set of points within one cell. As the cells are handled independently, this step can be executed in parallel on modern CPUs. In each iteration three (not colinear) points are selected randomly within the local set. Then a plane through these points is calculated and the number of inliers are counted (points in the cell which lie within a certain distance threshold to this plane). This number of supporting points is the maximization criterion: the calculated plane is used as best plane when its number of supporting points is bigger than the one of the best plane calculated during the previous iterations.

Algorithm 5.2 RANSAC algorithm for plane extraction

Given: a set of 3D distance measurement samples (data points)

Return: a set of planes approximating disjunctive subsets of the input points

Step 1: split the whole 3D scene into a regular grid of cubic cells

Step 2: assign the input points to the corresponding cells

Step 3: calculate fitting plane for each cell:

for every cell **do**

> find approximating plane using RANSAC algorithm (output: best fitting plane after certain number of RANSAC iterations)
>
> remove outliers with respect to the best RANSAC plane
>
> calculate least-square fitting plane for inliers

end for

Step 4: region growing – fuse matching planes of neighboring cells:

for every cell **do**

> compare plane parameters with those of all neighboring cells
>
> **if** the angle between plane normals is below the angular threshold and the distance of the center of gravity of points of neighboring cell to the plane is below the distance threshold **then**
>
> > mark both cells as belonging to the same region
>
> **end if**

end for

for all regions **do**

> calculate best fitting plane for all points of cells belonging to the same region

end for

Step 5: remove small planes supported only by a small amount of cells

---

This procedure is motivated by the fact that the repeated random selection of points and plane approximation converges to a "good" fitting plane if such a plane exists at all (model valid) and the number of iterations is high enough.

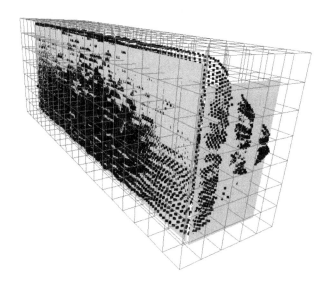

Figure 5.12 Splitting of a 3D point cloud into local sets via cubic grid cells

The plane fitting for all inliers within one cell as well as for all samples of one merged region is calculated via *principal component analysis*. A plane is described by its normal $\overrightarrow{n}$ and distance from origin d (equation 5.22).

$$< \overrightarrow{n}, \overrightarrow{x} > = d \tag{5.22}$$

According to [WGS03], the normal of the least-square fitting plane is the eigenvector of the smallest eigenvalue of covariance matrix A (equation 5.23).

$$A = \begin{pmatrix} \sum_{i=0}^{N} w_i x_i^2 & \sum_{i=0}^{N} w_i x_i y_i & \sum_{i=0}^{N} w_i x_i z_i \\ \sum_{i=0}^{N} w_i x_i y_i & \sum_{i=0}^{N} w_i y_i^2 & \sum_{i=0}^{N} w_i y_i z_i \\ \sum_{i=0}^{N} w_i x_i z_i & \sum_{i=0}^{N} w_i y_i z_i & \sum_{i=0}^{N} w_i z_i^2 \end{pmatrix} \tag{5.23}$$

Here, $x_i = x_i^{\mathrm{raw}} - \overline{x}$, $y_i = y_i^{\mathrm{raw}} - \overline{y}$ and $z_i = z_i^{\mathrm{raw}} - \overline{z}$ are input points centered around mean and weights w_i represent measurement uncertainty of the samples (set to 1 as uncertainty is not yet taken into account). d is given by equation 5.24

$$d = < \overrightarrow{cog}, \overrightarrow{n} > \tag{5.24}$$

where $\overrightarrow{cog} = (\overline{x}, \overline{y}, \overline{z})^T$ is the center of gravity of points belonging to the fitted plane. As A is a square symmetric matrix, its eigenvalues can be computed efficiently via *singular value decomposition*.

Table 5.13 lists all adjustable parameters used by the described algorithm. One question is how to choose reasonable values for the size of cubic
cells and the minimum number of points within each cell needed for ex-
tracting stable features. Naturally, the density of 3D samples decreases with
increasing distance between sensor and object (see figure 5.14). Assuming
an angular scan resolution of $\alpha = 0.5°$ horizontally and vertically and a
distance between adjacent laser beams of e =10 cm, a regular grid of 3×3
samples fits within a cell of 20×20×20 cm^3. In this case the maximum distance between sensor and objects is $d = \frac{e}{\tan \alpha} \approx 11.5$ m. This example shows
that objects which are further away than ≈ 11.5 m from the sensor generate too few samples for reliable plane extraction. Consequently, cell size and
minimum number of samples have to be chosen depending on sensor setup,
environmental conditions and size of features of interest.

step	parameter	value
1	cell size	200 mm
3	min. amount of points in a cell to start RANSAC alg.	10
3	number of RANSAC iterations	50
3	max. point-to-plane distance for inliers	50 mm
4	angular threshold for normals of neighboring planes	15°
4	distance threshold for cog of one plane to neighb. plane	50 mm
5	min. amount of supporting cells for plane filtering	10

Figure 5.13 Parameters used for plane extraction

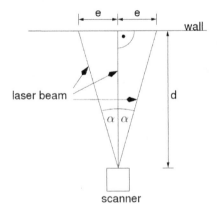

Figure 5.14 Distance between adjacent sampling points depends on distance between sensor and object

An example run of the described algorithm is shown in figure 5.10, with input data from figure 5.9. The scan resolution is 0.5° horizontally and 0.3° vertically and the number of samples is about 92000. The parameters have been chosen as shown in table 5.13. The plane extraction procedure took less than 2 seconds. For visualization, the resulting planes are clipped by the bounding boxes of their supporting samples. As expected, most stable features arise from floor and ceiling, followed by walls and cupboards. The window frames in the central part of the figure are represented by several small planes where region growing sometimes failed due to chosen plane orientation thresholds.

The output of the presented algorithm is a collection of 3D planes. At a higher level of environmental representation, these planes can now be used to extract semantically meaningful features such as the walls of a room and objects of furniture. However, corresponding strategies for this are beyond the scope of this book.

5.1.3   Feature-based metrical maps

Geometric invariants of a scan   At first, the regions of the map and the current scans which can be matched must be determined as the algorithms try to do partial matching. At least the extending information in the current scan cannot be part of the global map and therefore cannot be matched. The relative movement between two scans can be estimated using odometry, as the typical error does not accumulate over time if we do not try to estimate the global pose of the robot. This helps to identify the correct region. Another way of finding a good initial estimate is the usage of invariant attributes, that allow to compare two consecutive scans, despite to slight changes in position.

An example of invariants is the afore mentioned **box feature**. Slight variations in furniture or in the position of the vehicle in the room do not change the boxes seen. Another invariant is the **center of gravity** of a scan. Slight variations in the position and a turning at the same place will not change the center of gravity.

Let (x_i, y_i) be a scan taken in an robot centered coordinate system, then the center of gravity is given by

$$x_s = 1/N \sum_N^{i=1} x_i \tag{5.25}$$

$$y_s = 1/N \sum_N^{i=1} y_i \tag{5.26}$$

The center of gravity is an example of the concept of general **moments** $M_{p,q} = 1/N \sum_N^{i=1} (x_i)^p (y_i)^q$. For instance $M_{10} = x_s$ and $M_{01} = y_s$. Second order moments are the **moments of inertia** with respect to an axis under angle ψ through the center of gravity (x_s, y_s): Let d_i be the distance to the axis from point P_i with coordinates (x_i, y_i) then

$$f(\psi) = \sum_{i=1}^{N} (d_i)^2 \tag{5.27}$$

$\Rightarrow$ The line through (x_s, y_s) under an angle ψ is given by

$$y = m \cdot x + b \tag{5.28}$$

Then the distance of the line to the point (x_s, y_s) is calculated as shown:

$$y_s = m \cdot x_s + b \tag{5.29}$$

$$m = \tan \psi \tag{5.30}$$

$$b = y_s - m \cdot x_s \tag{5.31}$$

$$y_q = m \cdot x_i + b \tag{5.32}$$

$$\cos \psi = d_i / (y_i - y_q) \tag{5.33}$$

$$d_i = (y_i - y - q) \cdot \cos \psi \tag{5.34}$$

$$(d_i)^2 = (y_i - \tan \psi \cdot x_i - y_s - \tan \psi \cdot x_s)^2 \cdot \cos^2 \psi \tag{5.35}$$

Figure 5.15 shows the relation.

From here the moment of inertia through (x_s, y_s) with respect to angle ψ is

$$f(\psi) = \sum_{i=1}^{N} (d_i)^2 \tag{5.36}$$

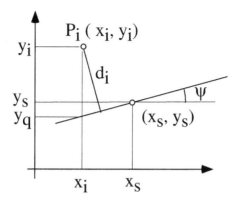

Figure 5.15 Distance with respect to the center of gravity and the angle *psi*

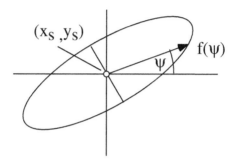

Figure 5.16 $f(\psi)$ forming the inertia ellipsis

Running through all angles ψ $f(\psi)$ forms an ellipsis, regardless of the given point cloud. The quotient of the length of the small and the long axis of the ellipsis as well as their direction are invariants of the point cloud as sketched in figure 5.16.

As the influence of points far away from the center of gravity is rather strong, a smoothing function is introduced: let d_0 be the medium distance to the center of gravity and d_i be calculated as above. Then d_i is smoothed by a function as shown in figure 5.17.

Here, a parameter α defines the steepness of the function described by

$$d_0 = \frac{1}{N} \cdot \sum_{i=1}^{N} \sqrt{(x_i - x_s)^2 + (y_i - y_s)^2} \tag{5.37}$$

$$d_i := \frac{d_i}{1 + \exp\left(\alpha \cdot (d_i - d_0)/d_0\right)} \tag{5.38}$$

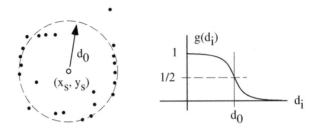

Figure 5.17 Left: center of gravity in a scan and medium distance of scan points right: a smoothing function, a parameter α describes the steepness of the function

While analyzing the data gathered during a scan of a laser distance sensor, one is able to distinguish four different kind of features as depicted in figure 5.18:

1. false edges: a sudden change in distances but no real edge

2. limit wall/ round edge

3. edge between two walls

4. real jump edge

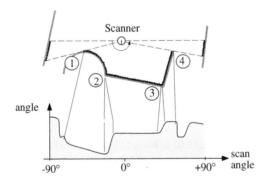

Figure 5.18 Different types of anchor points

Other than that, virtual edges occur. They can be found by applying a histogram based criterion. First of all, the scan image is rotated according to the main preferential orientation. Afterwards histograms are generated based on the scan data as illustrated in figure 5.19. Hence one is now able to determine the virtual intersection points of wall segments. The edge extraction process is presented in figure 5.20.

Using a scan (r_i, ϕ_i), segmentation can be performed according to algorithm 5.3, assuming $\Delta r_i = r_{i+1} - r_i$. Feature numbers 2 through 4 as well as virtual edges serve as anchor points of a scan. The distance of two anchor points P_i and P_j (with their coordinates (r_i, ϕ_i) and respectively (r_j, ϕ_j)) can be determined as shown in equation 5.39.

$$d_{ij} = \sqrt{(r_i)^2 + (r_j)^2 - 2r_i r_j \cos(\phi_j - \phi_i)} \qquad (5.39)$$

These distances are invariants of a scene.

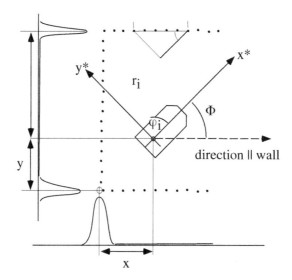

Figure 5.19 Anchor points from wall edges

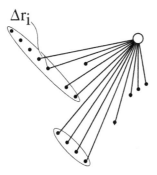

Figure 5.20 Finding edges in a radar scan

Algorithm 5.3 Segmentation

---

for $i = 1$ to N **do**
   if $|\Delta r_i| \leq s$ **then**
      form segment of (r_i, ϕ_i) and (r_{i+1}, ϕ_{i+1})
      $i := i + 1$
   else if $|\Delta r_i| > s$ and $|(r_{i+2} - r_i)| \leq s$ **then**
      add (r_{i+2}, ϕ_{i+2}) to segment
      mark (r_{i+1}, ϕ_{i+1}) as singular point
      $i := i + 2$
   else if $|\Delta r_i| > s$ and $|\delta r_{i+1}| < s$ **then**
      start new segment for (r_{i+1}, ϕ_{i+1})
      mark edge at (r_i, ϕ_i)
      $i := i + 1$
   else if $|\Delta r_i| > s$ and $|\delta r_i| > s$ **then**
      false edge detected
      mark (r_{i+1}, ϕ_{i+1}) is a singular point
      $i := i + 2$
   else
      mark (r_i, ϕ_i) is a singular point
      $i := i + 1$
   end if
end for

Extraction of anchor point type Let us assume the segments $\alpha(\phi_i)$ are given. Parts that are parallel to the ϕ-axis correspond with parts with a constant angle with respect to the orientation of the vehicle. Traversing from one parallel part to another, one denotes an edge (type 3). Straight parts not parallel to the ϕ-axis denote curved parts in the scene. Going from one part to another denotes a (type 2) edge. A discontinuity in the r_i-values denotes a real jump-edge (type 4). Hence, the scan can be transformed into a graph with vertices annotated with the type of edge and the edges with the distances as shown in figure 5.21 taken from [Web02]. The graph at hand is a compact description of a scene, invariant to slight changes in position and the orientation of the vehicle. In order to find the location of the vehicle, a graph corresponding to the scene just captured can be matched with graphs generated earlier. This approach remains within manageable dimensions as long as there are only a few graphs to be compared. This is the case if the vehicle position is approximately known.

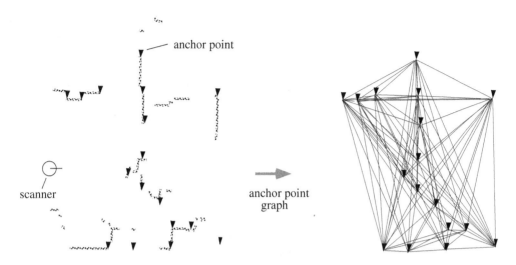

Figure 5.21 Generation of an Anchor Point graph (AP-graph) based on laser scanner measurements

Graph comparison

As one can see in figure 5.21, the anchor points found earlier can be connected in order to form a complete graph. This graph does not simply include the anchor points, but rather their relation is expressed by the help of edges that contain distance information as well. Assume two complete graphs G and $G*$ of a scene as given. Let us further assume that G features n anchor points and $n(n-1)$ distances and $G*$ contains m anchor points and $m(m-1)$ distances. The comparison of the two graphs can thus be reduced to a subgraph matching problem.

The momentary scan may involve 'ghost edges': Those can be either temporary object edges of obstacles (rather rare, only occurring with large obstacles) or the more common case of real edges that are obscured by obstacles.

In general only partial matching can be performed. In order to still be able to assign corresponding scenes, a particular search beginning with large distances in a data bank has to be implemented. Hence, if the number of found matches exceeds a preset threshold, the scene can be assumed as recognized.

Feature matching, continuous values Let there be two scenes with invariant features $M_i, \ldots, M_k$ and $M_i^*, \ldots, M_k^*$. Further, let the range of possible values of each feature M_i be continuous and thus $M_i \in [m_{ia}, m_{ie}]$. Therefore, scenes are described by feature vectors $\mathbf{M^T} = (M_1, \ldots, M_k)$ and

$\mathbf{M^{*T}} = (M_1^*, \ldots, M_k^*)$ interpreted as points in a room of k dimensions. A comparison is then based on the distance of the end points. In order to do that, the values have to be normalized $M_i \in (m_{ia}, m_{ie}) \rightarrow M_i \in [0,1]$:

$$M_i \longrightarrow \frac{M_i - m_{ia}}{m_{ie} - m_{ia}} \tag{5.40}$$

This assures all features are treated equally and the vector end-points define points in a k-dimensional unit cube. The Euclidean distance between M and M^* can be denoted as

$$d = \sqrt{\sum_{i=1}^{k} (M_i - M_i*)^2} \tag{5.41}$$

$$d \in [0, \sqrt{k}] \tag{5.42}$$

Feature matching, discrete range of values   Let us assume the existence of a discrete range of values for the features $M_i \in (m_{i1}, m_{i2}, \ldots, m_{in})$. If a relation $m_{i1} < m_{i2} < \cdots < m_{in}$ exists, then $m_{ir} \longrightarrow r$ and normalized $M_i \longrightarrow (M_i - 1)/(n - 1)$. If however this is not the case, only complete equivalence is considered: let $M_i = m_{ir}$ and $M_i^* = m_{is}$ then $d_i = (M_i - M_i^*) = (1 - \delta_{rs})$. If this equivalence is a condition sine qua non then

$$d = \delta_{rs} \cdot \sqrt{\sum_{j \neq i} (M_j - M_j^*)^2} + \sqrt{k} \cdot (1 - \delta_{rs}) \tag{5.43}$$

and otherwise $d = d_{\max}$.

5.2   Grid maps

Grid maps model the environment as a regular grid of cells with constant areas. The cells are filled with information extracted from noisy sensors. In order to produce consistent maps, this process requires either a known robot pose or a good pose estimate. The sensors used are predominantly ranging sensors like sonar or laser scanners. The lack of precise knowledge is expressed through various kinds of regions: high occupancy regions indicate objects, while lower values most likely represent free space. The advantage of this specific kind of map is the robustness and easy implementation. Their disadvantage is that they rely on exact pose estimation.

5.2.1 Occupancy grid maps

For many tasks it is useful to raster the environment. In general, a square raster with tiles of d cm width is used. Each tile gets a pair of indices (i,k), with d adjusted to the task at hand.

It is simple to build a grid map from laser radar distance readings. The dimension of a tile should be the resolution of the grid map. With $\Delta r \approx \pm 2.5$ cm and a radar cone of $0.5°$ up to a distance of $r = s \cdot 360/(2\pi \cdot 0.5) = 570$ cm a radar point occupies one tile and the cone denotes free space as shown in figure 5.22.

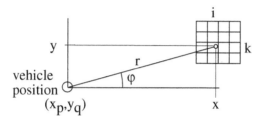

Figure 5.22 Building a grid map from laser radar measurements

Let $x = r \cos \varphi$ and $y = r \sin \varphi$ be the coordinates of the radar point and (x_p, y_q) the coordinates of the vehicle in a 2D scenario. The corresponding tile with index i,k and center (x_i, y_k) is given by

$$x_i - \frac{d}{2} < x + x_p \leq x_i + \frac{d}{2} \qquad x_i = i \cdot d \qquad (5.44)$$

$$y_k - \frac{d}{2} < y + y_q \leq y_k + \frac{d}{2} \qquad y_k = k \cdot d \qquad (5.45)$$

Each of the tiles described above is marked as free space (white) or obstacle (hatched) or partly free space (gray). Figure 5.23 shows a typical grid map with the mentioned parameters. If tiles are marked as occupied for vehicle navigation, there should remain a safety clearance of n of free space between the vehicle and the occupied region. E.g. suppose that $n = 7$ cm; with $d = n/\sqrt{2} \Rightarrow d = 5$ cm. In this case a grid map with 400 tiles/m^2 will be generated (for a field of 100 m$\times 100$ m, $4 \cdot 10^6$ tiles will be built up).

For a compact description of grid maps a **quadtree** representation can be applied. The environment is split recursively into blocks of $2^i \times 2^i$ tiles as shown in figure 5.24.

A node in the tree represents a $2^i \times 2^i$ square part of the environment. It is either a free space or represents an obstacle or is a mixture of both. If a node is marked as obstacle or free it will not be split any further. A mixed node will be recursively split in 4 equal areas, represented as 4 new nodes of the tree. If the whole environment is made up of $2^k \times 2^k$ tiles, then in general the quadtree will have much less than 2^{2k} nodes. The procedure is shown in algorithm 5.4.

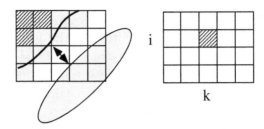

Figure 5.23 A typical grid map. White tiles represent free space, hatched tiles obstacles, and gray ones mixed spaces.

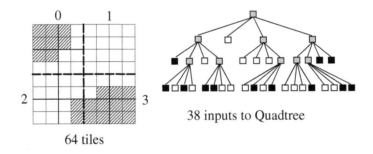

Figure 5.24 Quadtree representation of a grid map with 64 tiles.

Algorithm 5.4 Quadtree representation of a grid map

---

if $n = 0$ **then**
  it is a leaf node
else
  the map becomes the root node of a quad tree
end if
if the map is not uniformly colored **then**
  split the $2^n \times 2^n$ tiles into four maps of $2^{n-1} \times 2^{n-1}$ tiles each
  handle them the same way
else if a map is uniformly colored **then**
  it is a leaf node of the tree, denoting a block of $2^k \times 2^k$ tiles
end if

The concept may also be extended to three dimensions: splitting a 3D-space of $2^n \times 2^n \times 2^n$ voxels recursively into blocks of $2^k \times 2^k \times 2^k$. In figure 5.25 a 3D object is represented in a grid map.

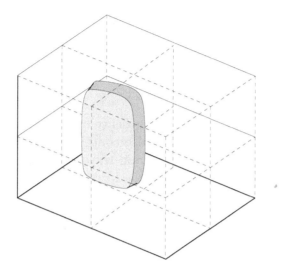

Figure 5.25 An object in a 3D grid

Probabilities Especially when using ultrasonic sensors with their rather unreliable measurements, it makes sense to include obstacle probability in the specification of the raster points. Performing more than one independent measurement, the probability of the presence of an obstacle at the specified location can be determined in a very simple manner. For each cell $C(i,j)$ the likelihood of occupancy can be computed as:

$$C(i,j) = \frac{\text{hits}(i,j)}{\text{hits}(i,j) + \text{miss}(i,j)} \qquad (5.46)$$

where $\text{hits}(i,j)$ represents the number of scans that indicated the presence of an obstacle at this cell while $\text{miss}(i,j)$ stands for the times the cell appeared to be empty. Thus the computation effort affiliated with e. g. Bayesian Filters can be reduced to the simple counter approach above.

5.3   Sector maps

A sector map widens the rigid structure of a grid map and allows to partition space in a more flexible way (see [AKSB07]). A sector map is divided up into

one or more *sectors*, hence its name. The most common separations consist
of uniform polar or rectangular sectors. Typically, a sector then contains the
following information:

- polar sector: angle and distance to the closest obstacle in the area the
  sector covers

- rectangular sector: x- and y-coordinates of the closest obstacle in the
  area the sector covers

While this is information about the presence of obstacles (and used for collision avoidance), a sector map can also contain other information, such as
the slope of the ground or overall terrain traversability.

In case of polar sector maps, space in front of an autonomous vehicle
is divided into sectors of some degree; e. g. $\Delta\varphi = 5°$ from $-120°$ to $+120°$.
Let the distance r_i to an obstacle be measured from the kinematic center at
an angle φ_i and the distance to the border of the vehicle be k_i. Then the
distance between vehicle and obstacle is $d_i = r_i - k_i$.

Figure 5.26 shows the sectors and figure 5.27 a typical sector map made
up of 24 sectors of $5°$ each, spanning $120°$.

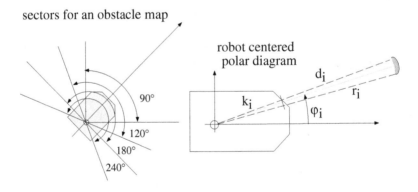

Figure 5.26 Typical angle range in front of the mobile robot, used for sector maps
(left). Obstacle representation inside a sector (right).

A sector map can be transferred into a grid map. Figure 5.27 shows a
typical sector map and its transfer to a robot-centered grid map. Its transformation into a world-centered grid map is presented in figure 5.28. In both
pictures free space and obstacles are marked in the tiles; blank tiles mark
unknown regions.

The transformation from a robot-centered sector map to world-centered
grid map is shown in figure 5.29.

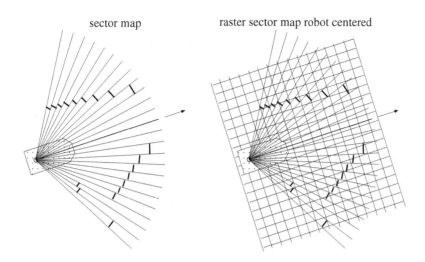

Figure 5.27 Transformation of a sector map to a grid map

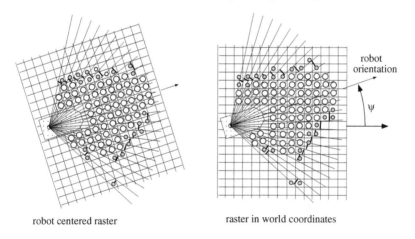

Figure 5.28 Robot-centered vs. world-centered grid map

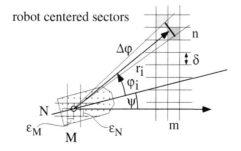

Figure 5.29 Transformation of a robot-centered sector map to a world-centered grid map

Let $\psi \neq 0$ be the orientation and $(x,y) = (M \cdot \delta + \varepsilon_M, N \cdot \delta + \varepsilon_N)$ be the position of the kinematic center of the vehicle in world coordinates. (M,N) is the index of the tile in which the kinematic center is located. ε_M and ε_N is the offset of the kinematic center in x- and y-direction due to the origin of the tile. δ is the edge length of the tiles. Suppose there is an obstacle at distance r_i in the sector i with an opening angle $\Delta\varphi$. Then all tiles with index (n,m), which fulfill one of the 3 pairs of inequations listed below have to be marked as obstacles and the space covered by the cone up r_i as free space.

$$(M + m) \cdot \delta - \varepsilon_M \leq r_i \cos(\varphi_i + \psi) \qquad \leq (M + m + 1)\delta - \varepsilon_M \quad (5.47)$$
$$(N + n) \cdot \delta - \varepsilon_N \leq r_i \sin(\varphi_i + \psi) \qquad \leq (N + n + 1)\delta - \varepsilon_N \quad (5.48)$$

$$(M + m) \cdot \delta - \varepsilon_M \leq r_i \cos(\varphi_i + \psi + \Delta\varphi/2) \leq (M + m + 1)\delta - \varepsilon_M \quad (5.49)$$
$$(N + n) \cdot \delta - \varepsilon_N \leq r_i \sin(\varphi_i + \psi + \Delta\varphi/2) \leq (N + n + 1)\delta - \varepsilon_N \quad (5.50)$$

$$(M + m) \cdot \delta - \varepsilon_M \leq r_i \cos(\varphi_i + \psi - \Delta\varphi/2) \leq (M + m + 1)\delta - \varepsilon_M \quad (5.51)$$
$$(N + n) \cdot \delta - \varepsilon_N \leq r_i \sin(\varphi_i + \psi - \Delta\varphi/2) \leq (N + n + 1)\delta - \varepsilon_N \quad (5.52)$$

For a sector up to 3 tiles may be marked as obstacles in the world-centered grid map.

A sector map has an arbitrary position and orientation in terms of the robot coordinate system. It is important to note that although the content of a sector map is usually generated by the data of a specific sensor, a sector map does not have to be aligned with this sensor. Instead, it can be placed anywhere in the area around the robot as a so-called "virtual sensor".

On the autonomous mobile robot ARTOS [AKSB07] of the TU Kaiserslautern sector maps are not only used for collision avoidance, but also as data source for an occupancy grid map built by the mapping system. The data of a laser range finder and two chains of ultrasonic sensors is put into three polar sector maps with uniform sectors. The two-step algorithm described in the following is used to extract the information from the sector maps and alter the occupancy counters of the grid map.

The first step of the algorithm deals with the grid cells whose counters have to be increased and the second one deals with the grid cells whose counters have to be decreased. The algorithm will be explained with the following example. Figure 5.30(a) depicts four obstacles o_1–o_4, the grid of a grid map and a sensor together with the sector boundaries of a sector map belonging to it.

In the first step, all sectors are traversed and the occupancy counters of the grid cells in which a sector's obstacle lies are incremented. The sector in which o_1 lies is not marked as occupied as o_1 is not in any of the sectors. The resulting grid map is shown in figure 5.30(b).

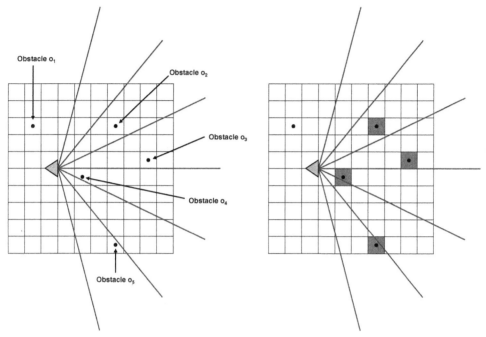

(a) Five obstacles lie in the area around the sensor. One of them, o_1, is currently not seen. The grid map is in its initial state.

(b) The cells in which visible obstacles lie have been marked as occupied.

Figure 5.30 First part of the creation of a grid map using a polar sector map as data source.

In the second step, the counters of the cells which are covered by a sector and are closer to the sensor than the obstacle in this sector are decremented. The problem is to determine the cells to which this applies. A comfortable way would be to have a list of these cells. But then the problem would be how to update this list when the robot's pose changes. So instead of calculating for each sector a list of the cells it covers, all cells in a user-specified rectangular area around the sensor are processed. For each of them the containing sector is determined. If the center of the grid cell is closer to the sensor than this sector's obstacle, its counter is decremented. Figure 5.31(a) shows an intermediate result of the map creation process, and figure 5.31(b) depicts the final grid map.

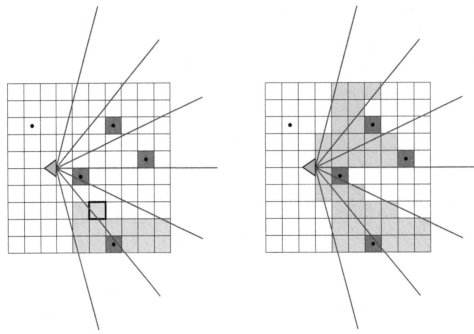

(a) The second step of the grid map creation algorithm is being executed. The cell with the bold frame has just been processed and marked as free.

(b) The second step of the grid map creation algorithm has been completed. All cells between the sensor and visible obstacles have been marked as free.

Figure 5.31 Second part of the creation of a grid map using a polar sector map as data source.

Note that in the second step, the counter decrementation is not limited to cells that are traversed by a sensor beam. Instead, the fact that the relevant obstacle in a sector is also the closest one is utilized to update the counters of more cells.

5.4 Topological maps

Compared to metrical maps, topological maps are more abstract descriptions of large-scale structures of the environment. Topological maps are typically represented as graphs in which navigation-relevant places are modeled as graph nodes and connections between places are indicated by graph edges. Often, some metrical information is also stored in a topological map, such as the coordinates of a place or the metrical length of the topological edge. Even if topological maps also contain such information, processing topological maps (generation, path finding) need less computation than metrical maps. Typical examples of topological maps are:

- bus lines and bus stops in a town,

- a highway network of a country,

- a network of stations and railway lines for a subway or railway system,

- a grid of the high voltage transmission lines of a country,

- the sewage system of a town.

Typical questions to be answered with help of a topological map are:

- Where do I change buses between stations A and B?

- Can I also drive from A to B via C or via D?

- How many stops are there on the way from A to B?

- If one transmission line fails, are there lines to circumvent the failed one?

- Where is the entrance to a main sewage line?

These questions are difficult to answer using geometric maps only. Therefore, for the solution of many mapping problems the transformation of a geometric map into a topological one is necessary, as shown in figure 5.32 and figure 5.33. In figure 5.32 three rooms are presented which are connected to each other by a passageway. Connecting the corners with edges which are not crossing any objects is shown in figure 5.33. This decomposes the map into regions free from objects.

This graph can easily be generated by visibility graph algorithms. The emerging regions can be represented by a graph as shown in figure 5.33. The nodes in this graph represent regions of the metrical map. The edges between nodes denote a passageway between the regions. This graph can be transformed into a **decomposition tree**. Here, all nodes belonging to one room are mapped to subgraphs; interconnection graphs represent the passageways between rooms. The decomposition tree algorithm divides the whole graph into subgraphs, in which nodes exist with one connection to another subgraph.

A possible interpretation of the generated **decomposition tree** is shown in figure 5.34.

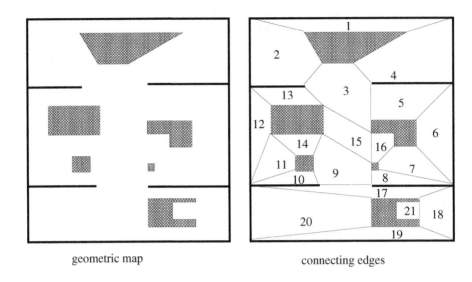

geometric map   connecting edges

Figure 5.32 Interconnecting edges in a map of 3 rooms

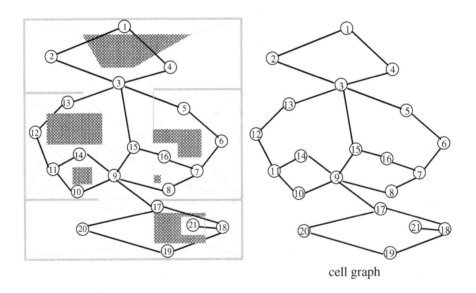

cell graph

Figure 5.33 Interconnecting regions and the corresponding graph

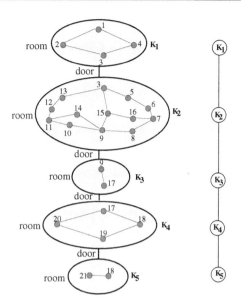

Figure 5.34 Decomposition tree

5.4.1 Growing neural gas net

Fritzke [Fri96] gave an algorithm to build a topological map of a scene from multiple measurements by different sensors.

Let the representation be a net with m nodes K_i made up from n-dimensional vectors $K_i = (b_{n-1,i}, \ldots, b_{0,i})$ with normalized components $b_{r,i} \in [0,1]$. The representation will be a net with m nodes K_i, $i = 1, \ldots, m$. In order to keep the number of nodes manageable, relevant nodes K_i will have to be calculated from many scene vectors $S = (s_{n-1}, \ldots, s_0)$. Each scene vector describes one set of normalized measured values $s_r \in [0,1]$. If the components of S are independent from each other – their covariance $c_{r,q} = \delta r, q$ – then the Euclidean distance between a node K_i and S is well defined:

$$\|K_i, S\| = \sqrt{\sum_{r=0}^{n-1} (b_{r,i} - s_r)^2} \tag{5.53}$$

Let a visiting counter z_m be attached to each node K_m and let D be a critical distance. Building up a growing neural gas net runs as shown in algorithm 5.5.

Having read in G scene vectors, then $G = \sum_{i=1}^{m} z_i$ and $m =$ number of nodes. There is still a difficulty to be solved: looking for the bmu in a large net given a scene vector S.

Algorithm 5.5 Growing Neural Gas Net

initialize: $m := 1$;   take a first measurement $K_m = S$; $z_m := 0$

repeat

  read S

  for all $j = 1, \ldots, m$ **do**

    $\|K_j, S\|$

  end for

  for all $j \neq i$ **do**

    $\|K_i, S\| \leq \|K_j, S\|$ {$- K_i$ is the node with minimum distance to S $-$ }

  end for

  if $\|K_i, S\| > D$ **then**

    {a new node is inserted}

    $m := m + 1$

    $K_m := S$

    $z_m := 0$

  else

    {$- K_i$ is the best matching unit $-$ **bmu** $-$ }

    if $z_i < Z$ **then**

      $z_i := z_i + 1$ {$-$ the visiting counter is increased $-$ }

      for all $r = 1, \ldots, n$ **do**

        $b_{r,i} := b_{r,i} + \varepsilon(s_r - b_{r,i})/z_i$ {$-$ the components of K_i are shifted into the direction of S while the weight of K_i increased $-$ }

      end for

    else if $z_i = Z$ **then**

      {$-$ the region around K_i is under represented and S gets a new node $-$ }

      $m := m + 1$; $z_i := Z/2$; $K_m := S$; $z_m := Z/2$

    end if

  end if

until FOREVER

The number of nodes to be checked should be kept manageable. The scene vectors as well as the nodes describe points in an n-dimensional unit cube. The direct approach as described in the algorithm to calculate $\|K_j, S\|$ for all $j = 1, \ldots, m$ and look for the minimum has an effort of $(2 \cdot m \cdot n)$. The idea is to preselect the nodes K_i, see figure 5.35.

- By multiplying the values of the components of the nodes by 2^p, the components of $K_i = (b_{(n-1),i}, \ldots, b_{0,i})$ become integer numbers $b_{r,i} \in (0, \ldots, 2^p - 1)$.

- Chop the interval $0, \ldots, 2^p - 1$ into 2^k parts t_q with $q = 0, \ldots, 2^k - 1$. The interval t_q contains the numbers $q2^{p-k} \leq b < (q+1)2^{p-k}$.

- The uppermost k bits of the values of b_r and $b_{r,i}$ form the indices q_r or $q_{r,i}$.

- Choose k such that $D = 2^{p-k}$.

- Form a list at index q in dimension r of all K_i with $(q-1)2^{p-k} \leq b_{ri} < (q+2)2^{p-k}$ under their number i. The list is ordered with respect to i. The storage effort is $(n2^k)$ lists with indices j and the values $b_{r,j}$.

- Candidates for the bmu are only those K_j whose number is to be found at **all** indices q_r, belonging to b_r.

- If the search fails, then S is a new node; there is no node K_j at a distance less than D around S.

- Let the indices $j_1, j_2, \ldots, j_l$ be found in all boxes q_r then only the distances from S to $K_{j1}, K_{j2}, \ldots, K_{jl}$ have to be calculated to find the minimum, the bmu.

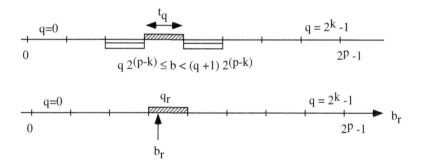

Figure 5.35 Splitting the realm of possible values

There are three parameters to be adapted to the scene at hand:

- The distance D, describing the influence of a node,

- the maximum value Z of a visiting counter z_j describing the number of matches before a split at the node K_j occurs,

- and the shift ε describing how much the current scene vector affects the bmu.

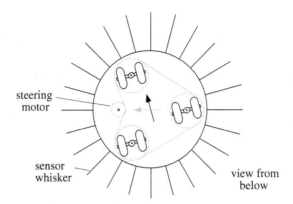

Figure 5.36 Sketch of ALICE

An example of building a topological map using a very simple vehicle, ALICE is described in [ZvP94]. Figure 5.36 shows the structure of the vehicle.

The robot has a circular shape with a diameter of 30 cm and is actuated by a synchro drive. ALICE is equipped with 24 photo sensors with an angle of beam spread of 15° each. The same number of touch sensors detect contact with walls as shown in figure 5.37.

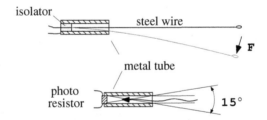

Figure 5.37 The touch sensor and the photo sensor of ALICE. The touch sensor is realized as a simple electrical switch. If the steel wire hits an obstacle the wire is bended and closes the contacts of the switch.

The measurements of 24 photo sensor $(I_0,\ldots,I_{23})$ are normalized to $\tilde{I}_i = I_i/I_{max} \in [0,1]$. The 24 touch sensor values $(T_1,\ldots,T_{24}) \in [0,1]$ are smoothed to cope with the rather large uncertainty of these sensors: $\tilde{T}_i = (2T_i+T_{i-1}+T_{i+1})/4, i = i \mod 24$. Only these 48 measurements describe the environment around ALICE. For the calculation of position and orientation of the vehicle the movement of the chain of the synchro drive is detected by light barriers and the wheel velocity measured by optical encoders.

The ALICE scene vector has 48 components. They provide input for a neural gas net modified slightly bit for the task of building a representation of the environment ALICE finds itself in.

- A first modification is to use the position information to build a graph: the vehicle actually has been driven from one best matching unit to the next, so there a line is drawn. The node is annotated with the position.

- The line is annotated with a visiting counter. It is used to cope with wrong measurements: once in a while a measurement gives wrong values. The sensor situation becomes unique and becomes a new best matching unit (**bmu**). From the last bmu to this new one a line is drawn and a visiting counter counted up. All the other counters attached to edges from that last bmu are diminished. It is highly improbable that a false measurement will be repeated. So the line will never be driven again. If the value of a visiting counter drops below the limit, the line is taken from the graph. This eliminates wrong measurements after a while.

- In order to get a reliable representation of the environment, the inevitable drift in direction has to be corrected. To this end the light impression of the photo sensors is used to correct a drift in direction.

Figure 5.38 shows a typical sensor situation of ALICE. Aside from the sensor values, the position and an identifier for the bmu part of the sensor situation. Figure 5.39 shows a test environment for ALICE and figure 5.40 a neural net formed after a while. Despite the primitive sensors, the environment can be recognized in form of the free space for ALICE.

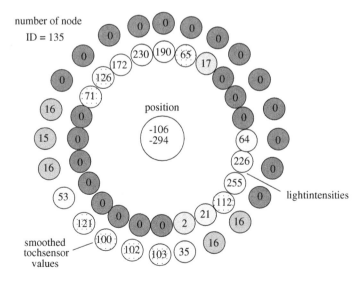

Figure 5.38 Example of sensor measurements (touch sensor values and light intensities) at an estimated position

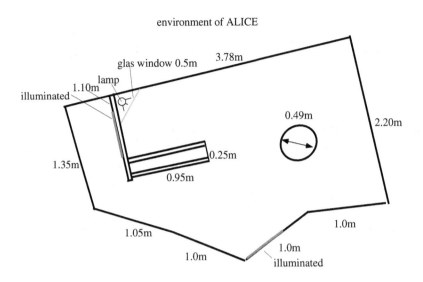

Figure 5.39   ALICE test environment

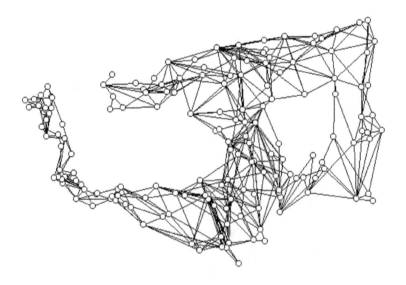

Figure 5.40   Topological graph of environment

5.5 Hybrid maps

As discussed before, metrical and topological mapping algorithms feature different characteristics (see figure 5.41). Hybrid approaches try to combine both types, allowing localization and map building with the high precision of metrical maps while retaining the compactness of topological maps.

	Topological Maps	Metrical Maps
Scale	Large-scale space	Small-scale space
Sensor inputs	Abstracts sensor inputs	Stores sensor inputs
Computational power	Low	High
Memory consumption	Low	High
Sensitive to noise	Less	More
Real-time mapping	Yes	Depends on computational power

Figure 5.41 Comparison of features of metrical and topological maps

Hybrid maps found in literature can be classified as two main types: abstraction-based and hierarchical hybrid maps. In abstraction-based hybrid maps, a metrical map of the environment is typically constructed as a basis, and an abstraction of the map is performed in order to create a compact topological representation. The benefit of this abstraction is more efficient planning of an approximate path to a given goal location than a detailed metrical map. However, the metrical map must often be kept for relocalization and obstacle avoidance purposes.

An example of a hybrid mapping strategy that derives a Voronoi-graph based topological map through abstraction of a long-lived, complete metrical map is presented in [Thr98]. Here, topological map building is initiated by thresholding an occupancy grid map built using Bayesian probability techniques. Then, the Voronoi diagram is built by selecting all free grid cells, with two nearest occupied grid cells (the *base points*) being equidistant. Cells on the Voronoi diagram are termed critical points if the base point distance is a local minimum. Lines between critical points and their base points divide metrical space into separate topological regions at locally narrow passages. From this regional decomposition, a topological graph can be generated. Figure 5.42 shows an example of the procedure.

With the help of this graph, fast path planning is possible without resorting to the detailed, underlying metrical map. During travel, however, the arrival at a topological node can only be detected by using localization techniques based on the detailed map, since the topological nodes are only characterized by their spatial position in the metrical map's frame of reference.

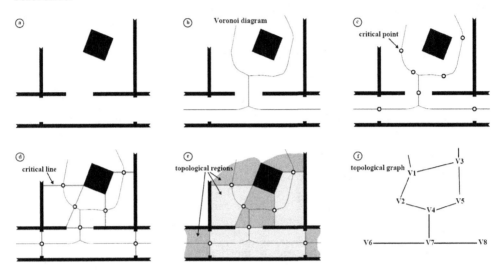

Figure 5.42 Steps in the topological abstraction scheme of [Thr98]. (a) Original thresholded occupancy map (b) Overlaid Voronoi diagram (c) critical points (d) critical lines (e) topological regions (f) resulting topological graph

Hierarchical hybrid maps on the other hand try to arrange the two map methodologies in a *hierarchical* fashion. This is accomplished by creating several local metrical maps with a limited scale and tying them together using a global topological map (figure 5.42). This approach uses the classical divide-and-conquer paradigm to address the scalability problems that are inherent in large metrical maps. Also, it prevents errors incurred during metrical mapping from spreading over the entire mapped area. However, one also has to pay a price for the segmented map structure, as information contained in partially overlapping local maps cannot be used to enforce global consistency. This can lead to increased uncertainty for each local map, especially if they are closely spaced.

In [TNS03], a hierarchical hybrid map of an indoor environment is presented which combines a global topological map with local metrical maps. On the global level, a topological map is constructed which contains intermediate *nodes* spanning the topological graph, leafs that signify metrical *places* and *corner lists* on the links between nodes (see figure 5.43). The corners can be extracted easily from the data of a 360° laser scanner and serve as landmarks for localization on the topological scale.

While the landmark-based localization strategy is used during robot travel along edges, the approach switches to a metrical method once the robot has arrived at a leaf node marked 'place'. The local map stored in this leaf is a line-based metrical map. For each line segment a 'line feature' is generated that is described by the angle of the perpendicular to the line and its length. The local feature map can be used for localization by modeling all feature parameters and the current robot pose as the state vector of an extended Kalman filter. With the use of this local metrical representation, the robot can travel freely in the vicinity of the leaf node, performing navigational tasks with high metrical precision.

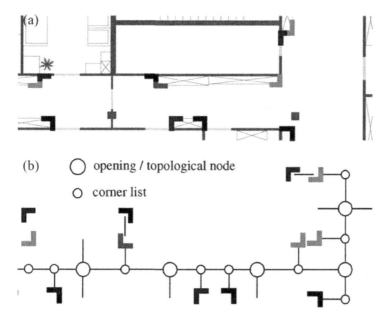

Figure 5.43 Hierarchical hybrid map of [TNS03]. (a) shows a portion of a hallway with extracted corner and opening features. (b) represents the resulting topological map with nodes and corner landmarks. Open arches either lead to metrical places or other hallways.

6 Simultaneous localization and mapping (SLAM)

In chapter 4 the localization problem is introduced, which is the estimation of the position and orientation of the AMR in its environment. It is shown that this problem can be solved with specific sensors or based on specific features of the environment. The selected features are those which could easily be detected by the robot sensor system. Additionally, chapter 5 describes different map generation techniques, for which a precise position and orientation of the robot is necessary.

For an autonomous system operating in unknown environment, a new challenge emerges: What needs to be done if the AMR is activated without precise information about its pose and no map of the environment exists? In this case the vehicle has to explore the surroundings, create a map, and track the pose. This somehow paradox situation is called the SLAM problem [SSC90, LDW91] (**S**imultaneous **L**ocalization **A**nd **M**apping).

6.1 The general approach

Suppose the AMR can extract features from its local environment that can be used as identifiers for landmarks. A precondition for the use of the landmark is the knowledge about its pose. The local configuration of these features (e. g. angle and distance due to the robot's coordinates system) yields a local feature map. For small movements it is easy to use these features to track the changing pose of the robot. In figure 6.1 (a) the robot measures the distances to three landmarks using e. g. a laser scanner. While moving, it can use its odometry to estimate the new pose as shown in figure 6.1 (b). This odometry-based robot pose could be used to estimate the new distances to the features extracted in figure 6.1 (a). Comparing the distances between robot and features with the measured ones, the AMR is able to correct its estimated pose as shown in figure 6.1 (c).

So far this is a localization with landmarks within the initial local feature map. By adding new features to this map, it can be incrementally extended until it covers the whole working space of the AMR. If one supposes that the distance measurement is absolutely precise and that enough of the old known features are observed in every extension step, then the SLAM problem is solved. In reality, we must assume that even after localization minimal errors in position and orientation persist due to the sensor systems used. These would not propagate through a pure relocalization process. If based on relocalization new features are added to the map however, the corresponding error results in wrong positions. Such errors accumulate during the whole map building process, resulting in unusable maps like the one shown in figure 6.2.

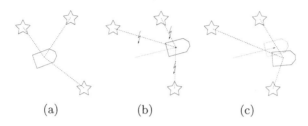

| (a) | (b) | (c) |

Figure 6.1 Relocalization relative to landmarks. In (a) the initial state with distances to three landmarks, in (b) the odometry-based pose with an error after some maneuvers and in (c) the corrected pose using the information from the distance sensors is shown.

 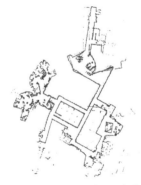

(a) A map resulting from pure integration of single scans

(b) Correct representation of the map

Figure 6.2 Pure integration of single scans from different poses leads to an unusable map.

The solution of this problem is depicted in figure 6.3. The main idea is to make small movements and stay within the range of as many known features as possible. Then new landmarks can be aligned properly. Figure 6.3 (a) is the situation at the end of the relocalization process (see also figure 6.1). In figure 6.3 (b) the vehicle has moved ahead and is properly relocated using two landmarks from the last local map. Because of the distance error when measuring the landmarks, the global vehicle pose is imprecise. Therefore, all currently observed features are displaced (the dashed landmarks show the correct poses). Via the relocalization process, the new landmarks on the right are properly located relative to the old ones. The displacement of the old landmarks can be estimated and used to align the new segment of the feature map. The resulting extended map is depicted in figure 6.3 (c).

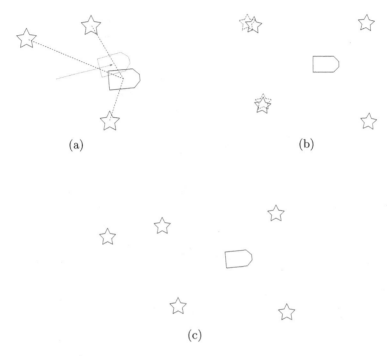

Figure 6.3   Relocalization relative to landmarks

This approach works if more than one old landmark is involved in the matching process. Thus, the best results are achieved by moving in small steps and performing updates frequently to extend the map only by a few (maybe just one) new landmarks.

6.2   Merging local maps

The example used above points out that one main problem of the SLAM approach is the merging of the feature maps. Using SLAM in indoor scenarios like that presented in figure 6.2 (b) usually involves line or point features that can be extracted by a 2D laser scanner. In the following two algorithms are presented that merge this type of feature maps.

6.2.1   Correlation of laser scans

Two laser scans are given, one from position (x_0,y_0,ψ_0), the other one from an unknown position (x,y,ψ) nearby. The setting is illustrated in figure 6.4. Let ψ_0 be the main direction in the scene. A scan from $P_0 = (x_0,y_0,\psi_0)$ is taken in a robot centered coordinate system as a point set $(\varphi_{0,i},r_{0,i})$, The same holds for a radar picture taken from $P_1 = (x_1,y_1,\psi)$ in the robot coordinate system centered at P_1. Both scans describe more or less the same scene. The calculation aimed to find the angular shift $\Delta\alpha$ and the lateral shifts Δx and Δy is performed in two steps:

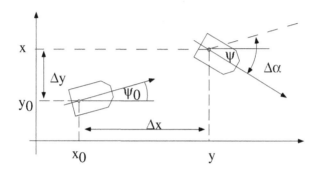

Figure 6.4 Two poses shifted by Δx and Δy

- Form the angle histograms (compare to figure 4.12) and correlate them with respect to turning by an angle. Let $\Delta\alpha$ be the angle with maximum correlation, then $\psi = \psi_0 + \Delta\alpha$.

- Correlate the angle adjusted scans with respect to lateral shifts (point histogram). Let Δx and Δy be the maxima in these correlations. The unknown position is then given by $x = x_0 + \Delta x$ and $y = y_0 + \Delta y$

The correlation of angle histograms is done as follows: Let $\{N_k\}_0$ and $\{N_k\}_1$ be two angle histograms over the full circle taken from positions P_0 and P_1 respectively with $(k - m) \mod K$. Here k counts the angular boxes of width $\delta\alpha$. Let the medians of these histograms be

$$\tilde{N}_0 = \frac{1}{K}\sum_{k=0}^{K-1} N_{k,0} \qquad (6.1)$$

$$\tilde{N}_1 = \frac{1}{K}\sum_{k=0}^{K-1} N_{k,1} \qquad (6.2)$$

The discrete correlation is

$$R_{0,1}(m) = \frac{1}{2K} \sum_{k=0}^{K-1} (N_{k,0} - \tilde{N}_0) \cdot (N_{k-m,1} - \tilde{N}_1) \qquad (6.3)$$

This is a function of the discrete shifts m of the angle histograms as shown in figure 6.5. There is one pronounced maximum between zeros at m_l and m_r. The maximum is described here by the center of gravity of the correlation values.

$$m_a = \frac{\displaystyle\sum_{m=m_l}^{m_r} R_{0,1}(m) \cdot m}{\displaystyle\sum_{m=m_l}^{m_r} R_{0,1}(m)} \qquad (6.4)$$

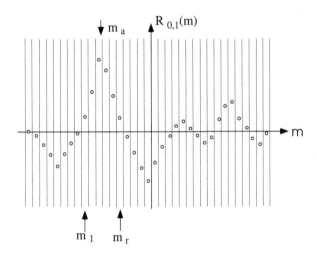

Figure 6.5 Correlations of angular histograms

The most probable angle of rotation is $\Delta\alpha = m_a \cdot \delta\gamma$. Turn the radar pictures:

- turn radar picture at P_0 with $\gamma_0 = \psi_0$ into the main axis direction of the scan as shown in figure 6.6, then radar point $(r_i, \varphi_i)_0 \longrightarrow (r_i, \phi_i - \gamma_0)_0$

$$x_{0,i} = r_{0,i} \cdot \cos(\varphi_{0,i} - \gamma_0) \qquad (6.5)$$

$$y_{0,i} = r_{0,i} \cdot \sin(\varphi_{0,i} - \gamma_0) \qquad (6.6)$$

- turn radar picture at P_1 with $\gamma_0 + \Delta\alpha$ into the same direction, as shown in figure 6.7 and radar point $(r_i, \varphi_i)_1 \longrightarrow (r_i, \varphi_i - \gamma)_1$

$$x_{1,i} = r_{1,i} \cdot \cos(\varphi_{1,i} - (\gamma_0 + \Delta\alpha)) \qquad (6.7)$$
$$y_{1,i} = r_{1,i} \cdot \sin(\varphi_{1,i} - (\gamma_0 + \Delta\alpha)) \qquad (6.8)$$

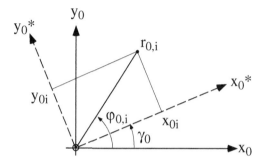

Figure 6.6 Turning P_0 by γ

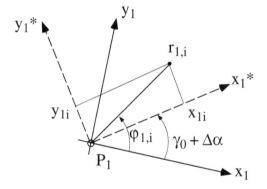

Figure 6.7 Turning P_1 by the same angle γ

The radar pictures form sets $\{x_{0,i}\}, \{y_{0,i}\}, \{x_{1,i}\}, \{y_{1,i}\}$ with $i = 0, \ldots, 719$. Their coordinate systems are aligned pointing into the same direction.

Now the point histograms may be calculated to find the lateral displacement: Take the set $\{x_{0,i}\}$, $i = 0, \ldots, 719$ and form boxes of width δx. Let $\delta x = 6$ cm be the resolution of the laser scanner taking the radar pictures. Let $P\delta x$ be the maximum detectable distance, thus outside $\pm P\delta x$ the boxes remain empty. Choose $N \approx 2P$. The histogram then has $2N$ boxes. Form a histogram $H_{0x} = \{h_{0xn}\}$ as shown in figure 6.8. Put all points with values between $n\delta x$ and $(n+1)\delta x$ into box h_{0,x_n}.

Do the same with points $y0, x1$ and $y1$ forming point histograms H_{0y}, H_{1x} and H_{1y}. The centers of gravity of these histograms can be expressed as:

$$\widetilde{h}_{0x} = \frac{1}{2P} \sum_{n=-P}^{+P} h_{0x_n} \tag{6.9}$$

$$\widetilde{h}_{0y} = \frac{1}{2P} \sum_{n=-P}^{+P} h_{0y_n} \tag{6.10}$$

$$\widetilde{h}_{1x} = \frac{1}{2P} \sum_{n=-P}^{+P} h_{1x_n} \tag{6.11}$$

$$\widetilde{h}_{1y} = \frac{1}{2P} \sum_{n=-P}^{+P} h_{1y_n} \tag{6.12}$$

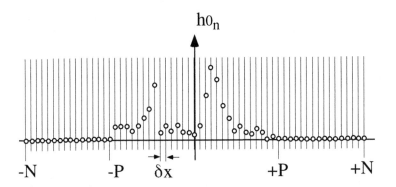

Figure 6.8 Point histogram

Afterwards H_{0x} with H_{1x} and H_{0y} with H_{1y} are correlated in x- and y-direction:

$$Qx_{01}(m) = \frac{1}{2P} \sum_{n=-P}^{+P} (h_{0x_n} - \widetilde{h}_{0x}) \cdot (h_{1x_{n-m}} - \widetilde{h}_{1x}) \tag{6.13}$$

$$Qy_{01}(m) = \frac{1}{2P} \sum_{n=-P}^{+P} (h_{0y_n} - \widetilde{h}_{0y}) \cdot (h_{1y_{n-m}} - \widetilde{h}_{1y}) \tag{6.14}$$

This procedure is repeated for all $-(N - P) \leq m < (N - P)$. The maxima m_x and m_y can be found in these correlations.

Then $\Delta x = m_x \cdot \delta x$ and $\Delta y = m_y \cdot \delta x$

$$\Longrightarrow P_1 = (\Delta x, \Delta y)_0 \qquad (6.15)$$

in a coordinate system $[x_0^*, y_0^*]$ centered at P_0. Figure 6.9 shows the result of these correlations.

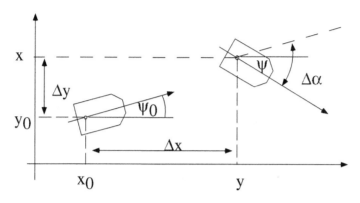

Figure 6.9   Correlating the Pose of P_0 to P_1

6.2.2   Correlation of point clouds

If the operational environment of the vehicle cannot be described with a set of line features, the above algorithm is not feasible. In this case it is often possible to extract point features of the environment (e.g. the point cloud of the laser scanner). [Bre04] describes a method which uses an adaption of the ICP algorithm [BM92] (see figure 6.10) and is able to compute the transformation between two point clouds.

(a)   (b)   (c)   (d)

Figure 6.10   The ICP algorithm. After building the correspondence pairs (a) the MSE is minimized (b). The resulting transformation allows to build new (better) correspondence pairs (c). The next minimization step leads to the final transformation which matches the two point sets correctly (d).

Given two point sets, M and D with $|M| = |D|$ that correspond (i.e. every point m_i matches with one point in d_i), the transformation from D

to M can be determined by minimizing the mean squared distances (MSE) between these corresponding pairs:

$$\text{MSE} = \frac{1}{|M|} \sum_{i=1}^{|M|} \|m_i - d_i\|^2 \qquad (6.16)$$

At first, the corresponding pairs are those which have the minimal Euclidean distance to each other. This is a search-problem (average time-complexity is $\mathcal{O}(n \log n)$) that can be solved using a kd-tree. The resulting corresponding pairs are not necessarily correct. The algorithm calculates a rotation and translation of the point cloud D in a way that equation 6.16 is minimized. After applying this transformation to the points in D, other pairs will be chosen as correspondence pairs. These two steps are repeated until the MSE falls below an error threshold or a fixed number of iterations are executed.

To calculate the transformation for the minimization of equation 6.16 the following method can be applied: The correspondence of m_i and d_i can be expressed by

$$m_i = \mathbf{R}d_i + \mathbf{T} + \varepsilon_i , \qquad (6.17)$$

with $\mathbf{R}$ being a rotation matrix, $\mathbf{T}$ being a translation vector and ε_i being a noise vector.

The vector ε_i reflects that the point clouds will not completely match because of sensor measurement errors. The optimal transformation $(\hat{\mathbf{R}}, \hat{\mathbf{T}})$ maps D to M while minimizing the MSE. Therefore, is can be written as

$$\text{MSE} = \frac{1}{|M|} \sum_{i=1}^{|M|} \left\| m_i - \hat{\mathbf{R}}d_i - \hat{\mathbf{T}} \right\|^2 \qquad (6.18)$$

The first step is to compute the rotation matrix $\hat{\mathbf{R}}$. Thus, the point sets should have the same centroid, which can be calculated by subtracting the center of gravity of the whole set from each point.

$$\bar{x} = \frac{1}{|X|} \sum_{i=1}^{|X|} x_i \qquad \bar{x}_i = x_i - \bar{x} \qquad (6.19)$$

Equation 6.18 can be rewritten and reduced:

$$\text{MSE} = \frac{1}{|M|} \sum_{i=1}^{|M|} \left\| \bar{m}_i - \hat{\mathbf{R}}\bar{d}_i \right\|^2 \qquad (6.20)$$

$$= \frac{1}{|M|} \sum_{i=1}^{|M|} \left(\bar{m}_i^T \bar{m}_i + \bar{d}_i^T \bar{d}_i - 2\bar{m}_i^T \hat{\mathbf{R}} \bar{d}_i \right) \qquad (6.21)$$

To minimize this equation, the last term must be maximized. This can be done by exploiting properties of unit quaternions [Hor87] or using a SVD[1]-based approach [AHB87] as described in the following.

For all real n-dimensional vectors a and b the following relationship between scalar products and traces of outer products is true:

$$a^T b = \sum_{i=1}^{n} a_i b_i = \mathrm{tr} \begin{pmatrix} b_1 a_1 & \cdots & b_1 a_n \\ \vdots & \ddots & \vdots \\ b_n a_1 & \cdots & b_n a_n \end{pmatrix} = \mathrm{tr}\left(ba^T\right) \qquad (6.22)$$

Thus, maximizing $\sum_{i=1}^{|M|} \overline{m}_i^T \hat{\mathbf{R}} \overline{d}_i$ means maximizing

$$\mathrm{tr}\left(\sum_{i=1}^{|M|} \hat{\mathbf{R}} \overline{d}_i \overline{m}_i^T\right) = \mathrm{tr}\left(\hat{\mathbf{R}} \mathbf{H}\right), \qquad \mathbf{H} = \sum_{i=1}^{|M|} \overline{d}_i \overline{m}_i^T \qquad (6.23)$$

Now, as the singular value decomposition yields

$$\mathbf{H} = \mathbf{U} \mathbf{\Sigma} \mathbf{V}^T \qquad (6.24)$$

with the orthonormal matrices $\mathbf{U}$ and $\mathbf{V}$ and a non-negative diagonal matrix $\mathbf{\Sigma}$, defining $\mathbf{X} = \mathbf{V} \mathbf{U}^T$ leads to a symmetric positive definite matrix:

$$\mathbf{X}\mathbf{H} = \mathbf{V}\mathbf{U}^T\mathbf{U}\mathbf{\Sigma}\mathbf{V}^T \qquad (6.25)$$
$$= \mathbf{V}\mathbf{U}\mathbf{\Sigma}\mathbf{V}^T \qquad (6.26)$$

For positive definite matrices the Cholesky decomposition yields

$$\mathbf{X}\mathbf{H} = \mathbf{A}\mathbf{A}^T \qquad (6.27)$$

and as the scalar product of two vectors has the largest positive value, if these vectors point into the right direction for every orthonormal matrix $\mathbf{B}$ it is true that:

$$\mathrm{tr}\left(\mathbf{A}\mathbf{A}^T\right) = \sum a_i^T a_i \geq \sum a_i^T \mathbf{B} a_i = \mathrm{tr}\left(\mathbf{B}\mathbf{A}\mathbf{A}^T\right) \qquad (6.28)$$

with a_i being the i-th column of $\mathbf{A}$. That means $\mathrm{tr}\left(\mathbf{X}\mathbf{H}\right) \geq \mathrm{tr}\left(\mathbf{B}\mathbf{X}\mathbf{H}\right)$ is true for every possible rotation matrix $\mathbf{B}$, and as $\mathbf{X}$ already is the product of two orthonormal matrices it solves the problem:

$$\hat{\mathbf{R}} = \mathbf{X} = V U^T \qquad (6.29)$$

---

[1] singular value decomposition

After aligning the rotation, the optimal translation aligns the centroids of both point sets:

$$\hat{\mathbf{T}} \quad = \quad \overline{m} - \hat{\mathbf{R}}\overline{d} \tag{6.30}$$

Because this algorithm tries to match each point in D with one point in M, outliers (i. e. points that do not correspond) must be detected and filtered before the algorithm is applied.

An example for the detection of outliers for the use of two-dimensional laser range finders is mentioned in [Bre04]. A scan S is a set of measurements $s_i = (r_i, \alpha_i)$ in the form of polar coordinates with fixed angular resolution $\Delta\alpha$:

$$S = \{s_i = (r_i, \alpha_i) \mid 0 \leq i < n\}, n = 360/\Delta\alpha \tag{6.31}$$

The single measurements can be ordered by their angles:

$$\alpha_i < \alpha_k \Leftrightarrow i < k \tag{6.32}$$

This allows for the use of a projection filter (see figure 6.11) to decide which scan points are visible from both locations and which are only visible from one location.

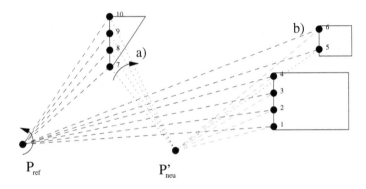

Figure 6.11 The projection filter. Only points 1–4 and 6 and are visible to P'_{new}. The others have to be purged using the proposed criteria.

This is important to reduce both scans to the possible correspondence set. The reference scan S_{ref} was taken from the position P_{ref}. A second scan that should be matched against S, was taken from the estimated position P'_{new}. The projection filter checks which scan points of S are visible from P'_{new}. Other points cannot be matched and have to be ignored by the ICP algorithm. This check is rather simple:

(a) Use equation 6.32 to determine points with reversed order. These are on the faces that pointed towards P_{ref} but point backwards for P'_{new}.

(b) In case of multiple points on one straight line connecting a point of S with P'_{new}, only the first point (closest distance to P'_{new}) can be visible. Little deviations of this line must be taken into consideration.

An example for the application of the SLAM algorithm utilizing ICP is shown in figure 6.12, where an AMR drives along a corridor, entering a room with maybe a locker on the right and a table with a chair on the left. Then it passes a second door and moves into another room.

Figure 6.12 An example for the application of a ICP-based SLAM algorithm

Figure 6.12 (a) shows the initial configuration of the robot (blue) and a second configuration after some maneuvers (red). The range of the laser scanner used is visualized by a distance circle. In this example the distance sensor can observe 360°. This is typically not the case. However, it allows larger movements and makes the example more compact. As one can see, the observation from one single pose is precise enough as mentioned in 6.1. However, the two configurations are misaligned. The first step is to use only those scan points located within the intersection of both distance circles. Any other point cannot be measured from both positions. Now the projection filter can be applied to the remaining points leaving only a few corresponding

candidates which can be matched by the ICP algorithm as can be seen in
(b). The resulting transformation aligns both scans, extending the map in
(c). After some more maneuvers a third scan is created in (d), reduced to
the corresponding information in (e) and aligned in (f). After adding two
additional scans the resulting map is shown in (m).

6.2.3   Loop closing

When implementing a pixel-based algorithm that registers point clouds and
merges them, it is important to not integrate them into one resulting global
map. That is because the relations between the successive measured local
pixel maps are lost, and therefore backpropagation of errors cannot be performed. To overcome this problem a dense topological map is introduced. It
contains the measured pixel maps as nodes. Two nodes that are connected
to each other represent two successive measurements. The relation between
these nodes is the estimated translation and rotation as described in the
previous section.

   One possible approach is the use of a circle around the position of the
map node added last. The square of the pose displacement (see section 6.1)
could be chosen as radius. If there is at least one other old node within this
circle, a loop candidate exists (see figure 6.13). The pixel map of each node
inside the circle must be compared with the pixel map of the new node. The
node which matches the new node best and has a correlation error below a
specific threshold closes a loop.

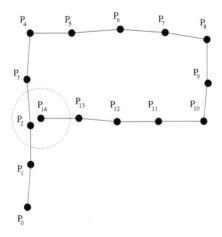

Figure 6.13 Loop closing problem. After 14 steps, the vehicle reaches a node with
a distance circle that contains the existing node P_2. It should be investigated for
closing the path.

6.3   Probabilistic methods

Current popular approaches to the SLAM problem are based upon stochastic observations. They take the positioning errors and noisy feature measurements into account and try to extract the positions of measured features and the robot's pose using an uncertainty model.

At the beginning of the process, the uncertainty model delivers fuzzy pose estimations. After each iteration a refinement of the model is performed, which leads to a precise map in the end.

6.3.1   An uncertainty model

To apply a probabilistic SLAM algorithm, first an uncertainty model of the robot's movement is required. With the robots pose $\hat{d}_i$ measured by odometry and the error $\tilde{d}_i$, the real position is $d_i = \hat{d}_i + \tilde{d}_i$. Then, the uncertainty model basically consists of the expectation μ_i and the variance σ_i^2:

$$\mu_i = \mathrm{E}\left(\tilde{d}_i\right) = \mathrm{E}\left(\left[d_i - \hat{d}_i\right]\right) \tag{6.33}$$

$$\sigma_i^2 = \mathrm{Var}\left(\tilde{d}_i\right) = \mathrm{E}\left(\left[d_i - \hat{d}_i\right]\left[d_i - \hat{d}_i\right]^T\right) \tag{6.34}$$

Typically the estimated pose of the robot is within an ellipsoid, as shown in figure 6.14.

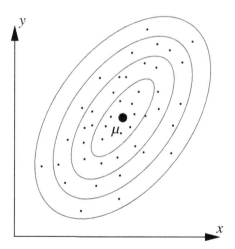

Figure 6.14 Typical uncertainty distribution for the robot position estimated by odometry

For vehicles with differential drive, the resulting distribution is as depicted in figure 2.20. There, a 2D ellipsoid is used for the position of the vehicle in a 2D environment. At the beginning of the robot movement straight ahead, the ellipsis is very small at the semi-minor axis because of the low slipping. Due to the bigger rotational error the semi-major axis grows faster. After the rotation at the corner, the ellipsis is turned as shown in figure 2.20. It can be observed that the semi-minor axis is not aligned with the straight forward direction of the robot movement. Depending on the drive characteristics of the vehicle, the distribution of the estimated pose may look different.

The same observations must be applied to the sensor measurements that detect the positions of the features. This leads to a sensor uncertainty model. In figure 6.15 the robot measures the distances to objects in its environment. In this example the distance sensor has a small angular error but a high distance error. The sensor uncertainty model results in elongate ellipses as shown in figure 6.15. The uncertainty models can be calculated using Kalman filters that can be found in section 4.7.

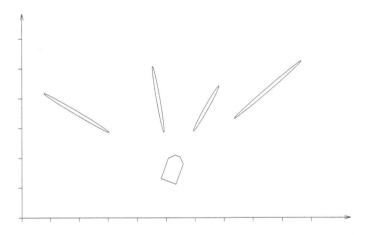

Figure 6.15 Uncertain distance measurements

6.3.2 SLAM as Bayesian network

FastSLAM [MTKW02] as a probabilistic approach describes a solution for the SLAM problem from a Bayesian point of view (see figure 6.16). FastSLAM factors the problem into the localization (i. e. the knowledge about the robot's path $s_1, s_2, \ldots, s_t$) and a collection of single landmark estimations θ_k that depend on the robot's estimated pose.

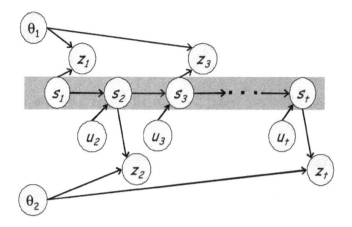

Figure 6.16 The SLAM problem as Bayesian network. The random variables are the position of the robot s_t, the control values u_t, the measured landmark positions z_t and the position of the landmarks θ_k. The directed edges represent condition dependencies. In the example the robot moves from s_1 to s_t with a sequence of control inputs $u_2, \ldots, u_t$. At the time $t = 1$ it observes landmark θ_1 via the measurement z_1. At $t = 2$ it observes θ_2 at z_2 and at $t = 3$ θ_1 again at z_3. It can be seen that the robot path depends only on the previous position and control input. (Image taken from [MTKW02])

In terms of the probabilistic approach, the robot's poses evolve according to the *motion model*:

$$p\left(s_t \mid u_t, s_{t-1}\right) \tag{6.35}$$

s_t is a probabilistic function of a control u_t and the previous pose s_{t-1}. The landmarks in the environment of the robot are characterized by their location, denoted as θ_k. By serializing the observation of multiple landmarks at the same time, sensor measurements for these landmarks are underlying the *measurement model*:

$$p\left(z_t \mid s_t, \theta, n_t\right) \tag{6.36}$$

θ is the set of all landmarks and n_t is the index of the landmark observed as z_t at the time t. To simplify the following description, the correspondence (value of n_t) is assumed to be known.

Given these two models, SLAM can be solved by determining the location of all landmarks and the robot poses based on measurements z_t and control inputs u_t:

$$p\left(s^t, \theta \mid z^t, u^t, n^t\right) \tag{6.37}$$

The superscript ^{t} describes a set of variables from time 1 to time t.

All individual landmark estimation problems are independent if the robot's path s^t and the correspondence n^t are known. So the rather difficult solution for (6.37) can be found by solving $k+1$ more simple problems:

$$p\left(s^t,\theta \mid z^t,u^t,n^t\right) \;=\; p\left(s^t \mid z^t,u^t,n^t\right) \prod_k p\left(\theta_k \mid s^t,z^t,u^t,n^t\right) \qquad (6.38)$$

6.3.3   The path estimator

FastSLAM implements a path estimator

$$p\left(s^t \mid z^t,u^t,n^t\right) \qquad (6.39)$$

using a particle filter that is similar to Monte Carlo Localization [TFBD01]: At each point in time, the algorithm maintains a set S_t of particles representing the posterior distribution $p\left(s^t \mid z^t,u^t,n^t\right)$. Each particle $s^{t,[m]}$ is considered as a guess of the robot's path, using the superscript notation $[m]$ to refer to the m-th particle in the set. Each particle set S_t is calculated incrementally from the set S_{t-1}, a control u_t and a measurement z_t. This is done by generating a temporary guess $s_t^{[m]}$ using the prior distribution $p\left(s_t \mid u_t, s_{t-1}^{[m]}\right)$. This basically means that the last guess together with the last control command are used to deduce the new guess (a similar process to the dead reckoning case). Assuming that S_{t-1} was distributed according to $p\left(s^{t-1} \mid z^{t-1},u^{t-1},n^{t-1}\right)$, the new set S_t is distributed according to $p\left(s^t \mid z^{t-1},u^t,n^{t-1}\right)$ as a proposal distribution. The new set S_t is then obtained by sampling from the temporary guesses with a probability that is proportional to an importance factor $w_t^{[m]}$. This results in a new distribution:

$$p\left(s^{t,[m]} \mid z^t,u^t,n^t\right) = w_t^{[m]} p\left(s^{t,[m]} \mid z^{t-1},u^t,n^{t-1}\right) \qquad (6.40)$$

In the following derivation, the conditional Bayes' theorem with the two events x,y and additional information e is applied

$$p\left(x \mid y,e\right) = \frac{p\left(y \mid x,e\right) p\left(x \mid e\right)}{p\left(y \mid e\right)} \qquad (6.41)$$

In addition, $z^t = z^{t-1} \cup z_t$ and $n^t = n^{t-1} \cup n_t$ is used to compute the weights $w_t^{[m]}$:

$$w_t^{[m]} \quad = \quad \frac{p\left(s^{t,[m]} \mid z^t, u^t, n^t\right)}{p\left(s^{t,[m]} \mid z^{t-1}, u^t, n^{t-1}\right)} \tag{6.42}$$

$$= \quad \frac{p\left(s^{t,[m]} \mid z_t, n_t, z^{t-1}, u^t, n^{t-1}\right)}{p\left(s^{t,[m]} \mid z^{t-1}, u^t, n^{t-1}\right)}$$

$$\overset{\text{Bayes}}{=} \quad \frac{\frac{p\left(z_t, n_t \mid s^{t,[m]}, z^{t-1}, u^t, n^{t-1}\right)}{p\left(z_t, n_t \mid z^{t-1}, u^t, n^{t-1}\right)} p\left(s^{t,[m]} \mid z^{t-1}, u^t, n^{t-1}\right)}{p\left(s^{t,[m]} \mid z^{t-1}, u^t, n^{t-1}\right)}$$

$$= \quad \frac{p\left(z_t, n_t \mid s^{t,[m]}, z^{t-1}, u^t, n^{t-1}\right)}{p\left(z_t, n_t \mid z^{t-1}, u^t, n^{t-1}\right)}$$

$$\propto \quad p\left(z_t, n_t \mid s^{t,[m]}, z^{t-1}, u^t, n^{t-1}\right)$$

$$\overset{\text{Total\_prob.}}{=} \quad \int p\left(z_t, n_t \mid \theta, s^{t,[m]}, z^{t-1}, u^t, n^{t-1}\right) p\left(\theta \mid s^{t,[m]}, z^{t-1}, u^t, n^{t-1}\right) d\theta$$

$$\overset{\text{Markov}}{=} \quad \int p\left(z_t, n_t \mid \theta, s^{t,[m]}\right) p\left(\theta \mid s^{t-1,[m]}, z^{t-1}, u^{t-1}, n^{t-1}\right) d\theta$$

$$= \quad \int p\left(z_t \mid \theta, s^{t,[m]}, n_t\right) p\left(n_t \mid \theta, s^{t,[m]}\right) p\left(\theta \mid s^{t-1,[m]}, z^{t-1}, u^{t-1}, n^{t-1}\right) d\theta$$

$$\propto \quad \int p\left(z_t \mid \theta, s^{t,[m]}, n_t\right) p\left(\theta \mid s^{t-1,[m]}, z^{t-1}, u^{t-1}, n^{t-1}\right) d\theta$$

$$= \quad \int p\left(z_t \mid \theta_{n_t}^{[m]}, s^{t,[m]}, n_t\right) p\left(\theta_{n_t}^{[m]}\right) d\theta_{n_t}^{[m]}$$

The last step assumes $p\left(n_t \mid \theta, s_t^{[m]}\right)$ being uniform and the landmark estimation relying on a Gaussian posterior $p\left(\theta_{n_t}^{[m]}\right)$, specified by the mean $\mu_{n_t}^{[m]}$ and covariance $\Sigma_{n_t}^{[m]}$ of the estimated position of θ_{n_t}. Now (6.42) can be solved in closed form.

6.3.4   The landmark estimators

The landmark estimators

$$p\left(\theta_k \mid s^t, z^t, u^t, n^t\right) \tag{6.43}$$

as the remaining part of equation 6.38 are implemented via Kalman filters. These estimators are conditioned on the robot pose, so each particle in S_t is extended by its own set of Kalman filters for the landmark estimators.

	Path	θ_1	θ_2		θ_k
1st Particle	s^t	μ_1,Σ_1	μ_2,Σ_2	$\cdots$	μ_k,Σ_k
2nd Particle	s^t	μ_1,Σ_1	μ_2,Σ_2	$\cdots$	μ_k,Σ_k
$\vdots$					
m^{th} Particle	s^t	μ_1,Σ_1	μ_2,Σ_2	$\cdots$	μ_k,Σ_k

Assume $n_t = k$. This means that the landmark θ_k is visible at time t and the estimation of $\theta_k^{[m]}$ can easily be obtained:

$$
\begin{aligned}
p\left(\theta_k \mid s^t,z^t,u^t,n^t\right) &= p\left(\theta_k \mid z_t,s^t,z^{t-1},u^t,n^t\right) \quad\quad (6.44)\\
&\overset{\text{Bayes}}{=} \frac{p\left(z_t \mid \theta_k,s^t,z^{t-1},u^t,n^t\right) p\left(\theta_k \mid s^t,z^{t-1},u^t,n^t\right)}{p\left(z_t \mid s^t,z^{t-1},u^t,n^t\right)}\\
&\propto p\left(z_t \mid \theta_k,s^t,z^{t-1},u^t,n^t\right) p\left(\theta_k \mid s^t,z^{t-1},u^t,n^t\right)\\
&\overset{\text{Markov}}{=} p\left(z_t \mid \theta_k,s_t,u_t,n_t\right) p\left(\theta_k \mid s^{t-1},z^{t-1},u^{t-1},n^{t-1}\right)
\end{aligned}
$$

For $n_t \neq k$, meaning landmark θ_k is not visible at time t, the distribution is not changed.

$$
p\left(\theta_k \mid s^t,z^t,u^t,n^t\right) = p\left(\theta_k \mid s^{t-1},z^{t-1},u^{t-1},n^{t-1}\right) \quad\quad (6.45)
$$

The updated equation (6.44) can be implemented using an extended Kalman filter, resulting in $\mathcal{O}\left(MK\right)$ computations for M particles and K landmarks per step t.

6.3.5   Numeric computation of FastSLAM

The following example with one landmark at position $\theta_1 = (28,24)$ is computed using a MATLAB implementation of Tim Bailey.[2] The vehicle starts its exploration in $(0,0,0)$ (position and orientation) and after the first movement it reaches $(0.67,0.03,0.00)$. As long as no landmark is measured, only the prediction step of the path estimator is evaluated to get one estimated position for each particle, as shown in table 6.1. As this example uses five particles, each has a weight of 0.2.

This way the particles integrate their position from odometry according to the underlying uncertainty model, until the first landmark θ_1 is measured at $t = 1$ with $z_1 = (29.34,0.38)$ (distance and angle). The true pose of the vehicle is $(7.45,2.96,0.38)$ and as up to this point no old landmark-estimations exist, no path-correction is performed. The new landmark estimation for each particle is the sum of the measurement and the estimated position:

---

[2] The SLAM package of T. Bailey is available at http://www.openslam.org

$$\mu_1^{[1]} = \begin{bmatrix} s_{1,1}^{[1]} + z_{1,1} \cos\left(s_{1,3}^{[1]} + z_{1,2}\right) \\ s_{1,2}^{[1]} + z_{1,1} \sin\left(s_{1,3}^{[1]} + z_{1,2}\right) \end{bmatrix} \tag{6.46}$$

$$= \begin{bmatrix} 7.54 + 29.34 \cdot \cos\left(0.37 + 0.38\right) \\ 2.94 + 29.34 \cdot \sin\left(0.37 + 0.38\right) \end{bmatrix}$$

$$= \begin{bmatrix} 28.93 \\ 23.03 \end{bmatrix}$$

Table 6.1 Particle states at $t = 0$

$[m]$	1	2	3	4	5
$w_0^{[m]}$	0.2	0.2	0.2	0.2	0.2
$s_0^{[m]}$	$\begin{bmatrix} 0.67 \\ 0.01 \\ 0.00 \end{bmatrix}$	$\begin{bmatrix} 0.68 \\ 0.04 \\ 0.01 \end{bmatrix}$	$\begin{bmatrix} 0.65 \\ 0.05 \\ 0.01 \end{bmatrix}$	$\begin{bmatrix} 0.67 \\ 0.04 \\ 0.01 \end{bmatrix}$	$\begin{bmatrix} 0.68 \\ 0.01 \\ 0.00 \end{bmatrix}$

In addition, the covariance of this landmark must be initialized. Therefore, the expected observation noise of the sensor, which consists of the average distance error $\sigma_1 = 0.1$ meters and average angular error $\sigma_2 = \frac{\pi}{180}$ radians ($1°$) must be known.

$$R = \begin{bmatrix} \sigma_1^2 & 0 \\ 0 & \sigma_2^2 \end{bmatrix}$$

Together with the observation matrix

$$G = \begin{bmatrix} \cos\left(s_{1,3}^{[1]} + z_{1,2}\right) & -z_{1,1} \sin\left(s_{1,3}^{[1]} + z_{1,2}\right) \\ \sin\left(s_{1,3}^{[1]} + z_{1,2}\right) & z_{1,1} \cos\left(s_{1,3}^{[1]} + z_{1,2}\right) \end{bmatrix}$$

the covariance can be derived as

$$\Sigma^{1,[1]} = GRG^T \tag{6.47}$$

$$= \begin{bmatrix} 0.13 & -0.13 \\ -0.13 & 0.14 \end{bmatrix}$$

The remaining particles can be computed analog (see table 6.2 and figure 6.17).

At $t = 2$ the same landmark θ_1 is measured again with $z_2 = (28.66, 0.39)$, but this time it is already known and the first update step can be executed. Therefore, the weights must be adjusted. The true pose of the vehicle is $(8.05, 3.26, 0.39)$. Together with the predicted pose in every particle, this can be used to predict the observation of θ_1 in this step for each particle:

$$d^{[m]} = \begin{bmatrix} \mu_{1,1}^{[m]} - s_{2,1}^{[m]} \\ \mu_{1,2}^{[m]} - s_{2,2}^{[m]} \end{bmatrix} \tag{6.48}$$

$$z_2^{[m]'} = \left(\left| d^{[m]} \right|, \mathrm{atan2}\left(d_2^{[m]}, d_1^{[m]} \right) - s_{2,3}^{[m]} \right) \tag{6.49}$$

Table 6.2 Particle states at $t = 1$

$[m]$	1	2	3	4	5
$w_1^{[m]}$	0.2	0.2	0.2	0.2	0.2
$s_1^{[m]}$	$\begin{bmatrix} 7.54 \\ 2.94 \\ 0.37 \end{bmatrix}$	$\begin{bmatrix} 7.56 \\ 2.83 \\ 0.36 \end{bmatrix}$	$\begin{bmatrix} 7.56 \\ 2.93 \\ 0.37 \end{bmatrix}$	$\begin{bmatrix} 7.57 \\ 2.91 \\ 0.36 \end{bmatrix}$	$\begin{bmatrix} 7.64 \\ 2.73 \\ 0.34 \end{bmatrix}$
$\mu_1^{[m]}$	$\begin{bmatrix} 28.93 \\ 23.03 \end{bmatrix}$	$\begin{bmatrix} 29.24 \\ 22.59 \end{bmatrix}$	$\begin{bmatrix} 28.99 \\ 22.98 \end{bmatrix}$	$\begin{bmatrix} 29.20 \\ 22.72 \end{bmatrix}$	$\begin{bmatrix} 29.60 \\ 22.19 \end{bmatrix}$
$\Sigma_1^{[m]}$	$\begin{bmatrix} 0.13 & -0.13 \\ -0.13 & 0.14 \end{bmatrix}$	$\begin{bmatrix} 0.12 & -0.13 \\ -0.13 & 0.15 \end{bmatrix}$	$\begin{bmatrix} 0.13 & -0.13 \\ -0.13 & 0.14 \end{bmatrix}$	$\begin{bmatrix} 0.13 & -0.13 \\ -0.13 & 0.15 \end{bmatrix}$	$\begin{bmatrix} 0.12 & -0.13 \\ -0.13 & 0.15 \end{bmatrix}$

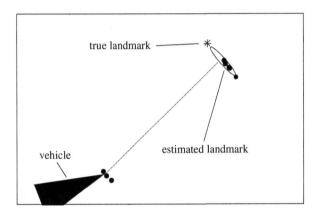

Figure 6.17 The first measurement of the landmark is observed. A model of the landmark is added to each particle.

To solve (6.42), the Jacobian matrices with respect to the vehicle state and the landmark state are needed:

$$H_{\text{vehicle}}^{[m]} = \begin{bmatrix} -\dfrac{d_1^{[m]}}{\left| d^{[m]} \right|} & -\dfrac{d_2^{[m]}}{\left| d^{[m]} \right|} & 0 \\ \dfrac{d_2^{[m]}}{\left| d^{[m]} \right|^2} & -\dfrac{d_1^{[m]}}{\left| d^{[m]} \right|^2} & -1 \end{bmatrix} \tag{6.50}$$

$$H_{\text{landmarks}}^{[m]} = \begin{bmatrix} \dfrac{d_1^{[m]}}{\left| d^{[m]} \right|} & \dfrac{d_2^{[m]}}{\left| d^{[m]} \right|} \\ -\dfrac{d_2^{[m]}}{\left| d^{[m]} \right|^2} & \dfrac{d_1^{[m]}}{\left| d^{[m]} \right|^2} \end{bmatrix} \tag{6.51}$$

Using these, the covariance matrix can be predicted, too.

$$\Sigma_2^{[m]'} = H_{\text{landmarks}}^{[m]} * \Sigma_1^{[m]} * H_{\text{landmarks}}^{[m]^T} + R \qquad (6.52)$$

Knowing the error with respect to the observation $\varepsilon^{[m]} = z_2 - z_2^{[m]'}$ the weights $w_2^{[m1]}$ can be computed:

$$w_2^{[m1]} = \frac{e^{-\frac{\varepsilon^{[m]^T} \Sigma_2^{[m]'-1} \varepsilon^{[m]}}{2}}}{2\pi \sqrt{\left|\Sigma_2^{[m]'}\right|}} \qquad (6.53)$$

The resulting weights are $(28.51, 32.63, 33.15, 24.86, 31.2)$ and must be normalized to sum up to one as can be seen in table 6.3.

The landmark estimations can be updated using a Kalman Filter with the prior state $\mu_1^{[m]}, \Sigma_1^{[m]}$, the innovation $\varepsilon^{[m]}, R$, and the linearized observation model $H_{\text{landmarks}}^{[m]}$ applying Cholesky factorization.

At last, the resampling is done which does not change anything in this step due to the still relative equal weights. They are all larger than $\frac{3}{4}$ of the average weight, which is chosen as limit to be kept.

The resulting values can be seen in table 6.3 and figure 6.18.

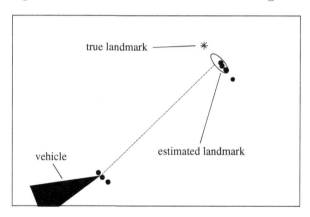

Figure 6.18 The next measurement can be used to weigh the particles and reduce the variance in the model of the landmark position.

At $t = 5$ of this example, the resampling part changes. Here, the weights are $(3.89, 6.13, 13.97, 3.91, 13.18)$ with an average of 8.22. So three of the particles are weighted below $\frac{3}{4}$ of the average weight and are resampled from the better ones. With five particles and the remaining weights it is easy to see that the third particle will be copied twice and the fifth particle once to fill the gaps. After this the weights are reset. The true position of the vehicle is $(9.86, 4.18, 0.42)$ and the result is shown in table 6.4 and figure 6.19.

Table 6.3 Particle states at $t = 2$

$[m]$	1	2	3	4	5
$w_2^{[m]}$	0.19	0.22	0.22	0.17	0.21
$s_2^{[m]}$	$\begin{bmatrix} 8.22 \\ 3.30 \\ 0.39 \end{bmatrix}$	$\begin{bmatrix} 8.22 \\ 3.17 \\ 0.37 \end{bmatrix}$	$\begin{bmatrix} 8.21 \\ 3.27 \\ 0.39 \end{bmatrix}$	$\begin{bmatrix} 8.22 \\ 3.27 \\ 0.38 \end{bmatrix}$	$\begin{bmatrix} 8.29 \\ 3.06 \\ 0.36 \end{bmatrix}$
$\mu_2^{[m]}$	$\begin{bmatrix} 28.73 \\ 23.28 \end{bmatrix}$	$\begin{bmatrix} 29.07 \\ 22.81 \end{bmatrix}$	$\begin{bmatrix} 28.81 \\ 23.18 \end{bmatrix}$	$\begin{bmatrix} 28.96 \\ 23.02 \end{bmatrix}$	$\begin{bmatrix} 29.4 \\ 22.43 \end{bmatrix}$
$\Sigma_2^{[m]}$	$\begin{bmatrix} 0.06 & -0.06 \\ -0.06 & 0.07 \end{bmatrix}$	$\begin{bmatrix} 0.06 & -0.06 \\ -0.06 & 0.07 \end{bmatrix}$	$\begin{bmatrix} 0.06 & -0.06 \\ -0.06 & 0.07 \end{bmatrix}$	$\begin{bmatrix} 0.06 & -0.06 \\ -0.06 & 0.07 \end{bmatrix}$	$\begin{bmatrix} 0.06 & -0.06 \\ -0.06 & 0.07 \end{bmatrix}$

Table 6.4 Particle states at $t = 5$

$[m]$	1	2	3	4	5
$w_5^{[m]}$	0.2	0.2	0.2	0.2	0.2
$s_5^{[m]}$	$\begin{bmatrix} 9.96 \\ 4.11 \\ 0.41 \end{bmatrix}$	$\begin{bmatrix} 9.96 \\ 4.11 \\ 0.41 \end{bmatrix}$	$\begin{bmatrix} 9.96 \\ 4.11 \\ 0.41 \end{bmatrix}$	$\begin{bmatrix} 10.08 \\ 3.83 \\ 0.38 \end{bmatrix}$	$\begin{bmatrix} 10.08 \\ 3.83 \\ 0.38 \end{bmatrix}$
$\mu_5^{[m]}$	$\begin{bmatrix} 28.54 \\ 23.52 \end{bmatrix}$	$\begin{bmatrix} 28.54 \\ 23.52 \end{bmatrix}$	$\begin{bmatrix} 28.54 \\ 23.52 \end{bmatrix}$	$\begin{bmatrix} 29.2 \\ 22.71 \end{bmatrix}$	$\begin{bmatrix} 29.2 \\ 22.71 \end{bmatrix}$
$\Sigma_5^{[m]}$	$\begin{bmatrix} 0.02 & -0.02 \\ -0.02 & 0.03 \end{bmatrix}$	$\begin{bmatrix} 0.02 & -0.02 \\ -0.02 & 0.03 \end{bmatrix}$	$\begin{bmatrix} 0.02 & -0.02 \\ -0.02 & 0.03 \end{bmatrix}$	$\begin{bmatrix} 0.02 & -0.02 \\ -0.02 & 0.03 \end{bmatrix}$	$\begin{bmatrix} 0.02 & -0.02 \\ -0.02 & 0.03 \end{bmatrix}$

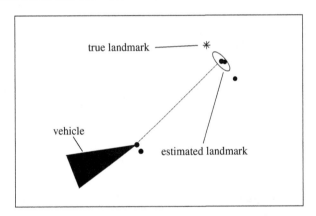

true landmark

vehicle

estimated landmark

Figure 6.19 Resampling removed the worst particles and copied the better ones.

After some more steps the relation between the landmark and the estimated vehicle are stabilized, even if they differ from the true positions. The map and the vehicle's pose within the map are consistent (figure 6.20).

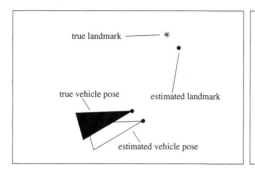 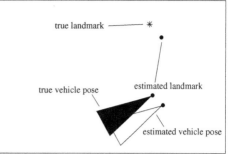

Figure 6.20 The algorithm created a consistent map.

6.4 Exploration of the environment

Task: The vehicle wakes up in an unknown environment, thus it has to explore. Method: Take crude scans of the environment, combining readings from ultrasound and laser distance sensors with an angular resolution of e. g. $\delta\varphi = 6°$ as shown in figure 6.21.

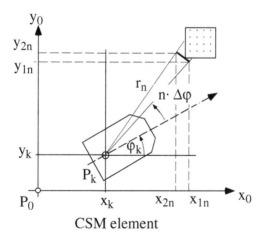

Figure 6.21 CSM element

These elements form the current sensor map (CSM) from the momentary position of the vehicle.

- Set a counter to $i = 0$

- Take a CSM. Label the current position P_i. Let there be possible passages at the perimeter of the scan. They are broad enough to let the vehicle pass through.

- Name the middle of these passages points of interest (POI) as shown in figure 6.22. Pack still unnamed POIs except one on a stack and set $i := i + 1$. Drive towards the middle of this one passage, label it P_i and take a new current sensor map from this point. Both CSMs will overlap and may be fused. From the current position P_i there is at least one way back to P_0.

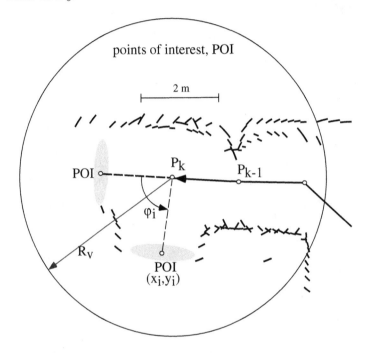

Figure 6.22 Points of Interest (POI)

- if in the CSM there is no unknown passage detected, take a POI from the stack and drive to it. Take a new CSM and set $i := i + 1$.

- when the stack is empty, the exploration is complete.

While driving through the environment, a representation in form of a graph consisting of POIs as nodes and driven ways as edges is built. Seeing a known POI P_j from the current point P_i, form an edge between both POIs instead of packing it onto the stack. This defines alternative paths through the environment.

There are two weak points in this algorithm:

- The vehicle must be able to detect the second floor in a park deck: it finds itself at exactly the same location but one floor upwards. Though the position is the same, the environment looks completely different. To circumvent this problem, the signal of an inclinometer must be integrated to show the height reached. The integrated error of the inclinometer signal must not be larger than $\pm 1m$ to decide wether the vehicle is still at the same level or one floor up/downwards. This problem increases for a vehicle running through a 3-dimensional sewage system.

- the odometry must be precise enough to detect the beginning of a large circular way as one already visited. This is difficult if the vehicle comes to the same location but from the other side. This is a general problem of map building that plagued the explorers of former times more than once. A possible solution is described in the next section.

The figures 6.23 and 6.24 show an experimental environment and a run through a large university building.

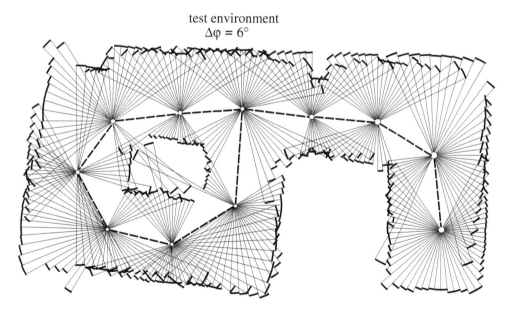

Figure 6.23 Test environment with CSMs

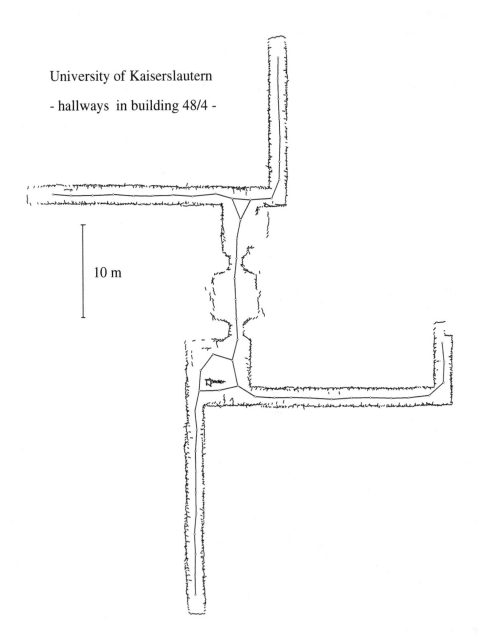

University of Kaiserslautern

- hallways in building 48/4 -

10 m

Figure 6.24 Experiment with CSMs in a test environment

7 Navigation

The aim of navigation is to drive the vehicle through its environment. This task splits into three different subtasks: The *global path planning* deals with finding a suitable path from a starting point to a goal point using a given representation of the environment. The *local path planning* defines path points taking into account the vehicle dimensions and kinematic constraints. *Path control* describes the task of generating suitable steering commands for following a precomputed path represented by reference points.

The following sections give more detailed information on these three subtasks and introduce typical application scenarios.

7.1 Global path planning

The aim of global path planning is to describe a path through the environment from a starting point S to a goal point Z, fulfilling some side conditions. One is to find the path with the lowest costs. This can depend on the length of the way, energy consumption for travelling, required time or a balanced mixture of that. Another condition can be that a path must be found to cover all free space, e. g. for cleaning vehicles.

The complete path is described by a set of points connecting subgoals. Often these points are nodes in a topological graph (i. e. in road maps: drive from A to B) with the details to be planned later on a geometric map.

The following sections describe three different scenarios. The first one is to plan a path minimizing the "costs" of the journey. The algorithm is the A*-algorithm, operating on a topological graph. The second one is solving a maze: you have a goal; find a path that ends at this goal. The third one is the task of finding back to the entrance of a maze.

7.1.1 A*-algorithm

Take an interconnected topological graph with nodes $K_1, \ldots, K_n$ and edges s_{ij} between nodes K_i and K_j. The edges are annotated with "prizes" p_{ij} which may be costs for a ticket, path lengths, number of narrow curves etc. The aim is to find the cheapest connection from a start node S to a goal node Z. Let an optimistic guess be annotated to each node K_i: the prize h_i to be payed for the direct route from $K_i \longrightarrow Z$. At Z the prize is $h = 0$. The total costs of a path $(S, K_i, K_j, \ldots, K_q, K_r)$ are $g = (p_{si} + p_{ij} + \cdots + p_{qr}) + h_r$ as sketched in figure 7.1. In 1968 Nilson [HN68] found the A*-algorithm to solve this task.

Figure 7.1 Costs of a path in a topological graph with annotated edges

Let W_r be the cheapest path from the starting point S to the goal Z up to node K_r. Let node K_q be the last node visited. Directly connected with node K_r apart from node K_q are nodes $K_{s1}, \ldots, K_{sn}$ with costs $g_{s1}, \ldots, g_{sn}$. Let the path via K_v be the cheapest one. K_v need not be one of the nodes $K_{s1}, \ldots, K_{sn}$, but may be the end point of another path already checked. Then develop node K_v further on. Memorize all paths and their costs which were already checked. Once $h = 0$ then the goal Z has been reached and a cost minimal path found.

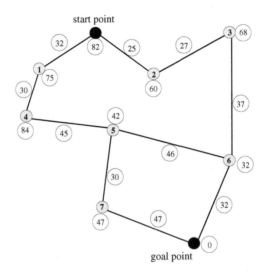

Figure 7.2 Exemplified graph used as an example for the A*-algorithm

In order to exemplify the algorithm, figure 7.2 shows an example of a topological graph. At each node the optimistic costs to the goal – in this case direct distance – are given and the prize p_{qr} at each edge – in this case the distance between the nodes. Figure 7.3 shows the steps in the development of the A*-algorithm. Notice the switching from one path to the next, once the costs of the former get higher: $(S,2,3)$ to $(S,1)$ and $(S,1,4)$ to $(S,2,3)$.

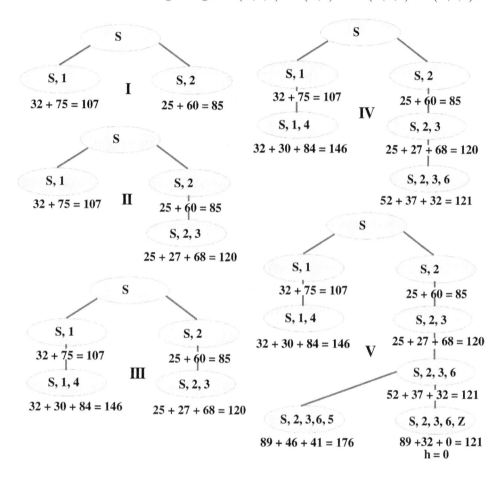

Figure 7.3   Development of A*-algorithm for the example given in figure 7.2

The implementation may describe the graph as an adjacency list with additional information regarding the prizes and an optimistic guess of the costs to reach the goal. Figure 7.4 shows this list for the example graph.

Aside from the adjacency list, a leaf list is formed as shown in figure 7.5, denoting the paths already checked. On this leaf list the algorithm develops: Pointers go from the momentary end node to the start node. Once $h = 0$, an optimal path has been found and the pointers in this list are reversed to point from the start to the goal. Figure 7.5 and figure 7.6 show the different steps in developing the solution.

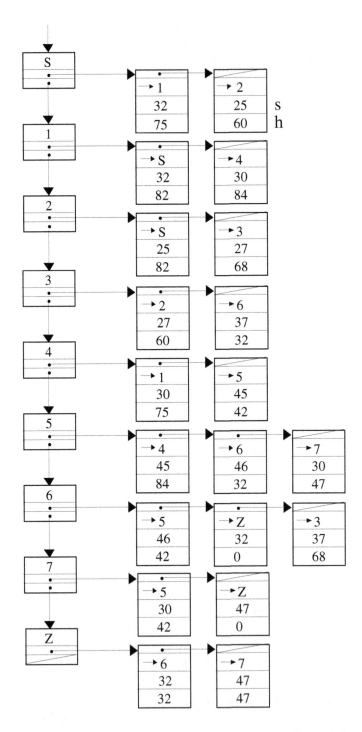

Figure 7.4 Adjacency list for the example graph given in figure 7.2

leaf_ lists ordered to costs

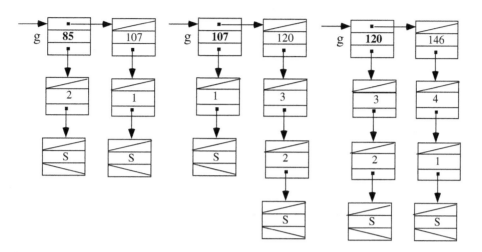

Figure 7.5 Leaf list 1

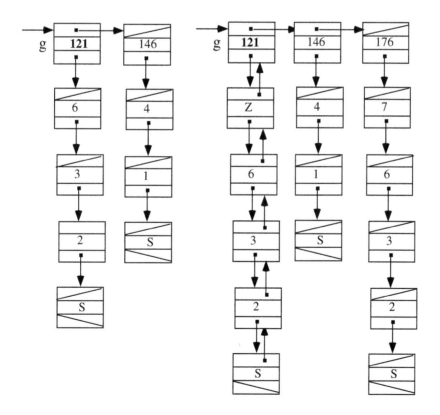

Figure 7.6 Development of the algorithm

7.1.2   Solving a maze

The classical algorithm to solve a maze is the **string of Ariadne** to find a
path to a hidden goal somewhere in a (3-dimensional) maze and back home
again without a map of the maze. The princess Ariadne of Crete offered a roll
of string to Theseus when he set out to find the Minotaur sitting somewhere
in a huge maze. Given a large ball of string and the ability to

- follow a wall

- turn back

- recognize junctions

- recognize the goal

- lay out a string and take it up again

- recognize a string on the ground

- follow a string to the next junction

the problem can be solved according to algorithm 7.1.

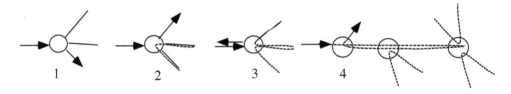

Figure 7.7   Using a String to Solve a Maze

For practical purposes, this algorithm is not very useful. But based on
this general scheme a very useful one is shown as follows.

7.1.3   Back tracking algorithm

Let the maze be represented by a topological graph as shown in figure 7.8,
described by an adjacency list as figure 7.9 shows.

Each node in the list gets a pointer to the adjacent nodes $K_j, \ldots, K_m$
and as additional information the length w_{ij} of the edge connecting nodes K_i
and K_j and a marker $m_{ij} \in [\texttt{nil},\texttt{free},\texttt{occupied},\texttt{visited}]$. Then the back
tracking runs as described in algorithm 7.2.

Algorithm 7.1 String of Ariadne

$(\star)$

if the goal has been recognized **then**

    turn; roll up the string again; follow the string up to the entrance

else

    {– the goal has not been recognized yet – }

    drive and lay out the string behind you until you reach a junction.

    At a junction check the exits

    if there is an exit to the right **then**

        if there is no string at this exit **then**

            follow this exit and go on at $(\star)$ {– figure 7.7 (1) – }

        else if you find a string there already **then**

            look for the first free exit to the left {– figure 7.7 (2) – }

            follow this exit and go on at $(\star)$

        else

            {you do not find an exit without a string}

            turn back; follow the string to the next junction {– figure 7.7 (3)

            and (4) – }

            go on at $(\star)$

        end if

    end if

end if

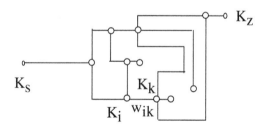

Figure 7.8 A maze as a topological graph

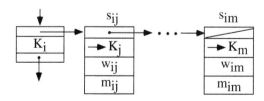

Figure 7.9 The adjacency list for the maze

Algorithm 7.2 Back tracking algorithm

---

initialization: $i := a; W := 0; m_{ik} =$ nil

for all edges **do**

$(\star)$

if $K_i = K_Z$ **then**

    {– goal found – }

    $w =$ path length to goal $\diamond$

else

    in node K_i look at all n edges $s_{ij}, \dots, s_{im}$ beginning at K_i;

    {– K_i itself was reached from node K_h – }

    if $m_{ij} =$ nil **then**

        {– it was the first visit at K_i – };

        $m_{ij} :=$ visited

        if there are edges going out from K_i marked nil **then**

            mark them as free;take an edge s_{ik};

            mark it $m_{ik} :=$ visited and drive to node K_k;

            $w := w + w_{ik}$; $i := k$; go to $(\star)$

        else

            {– a leaf of the graph was detected – }

            $m_{ij} :=$ occupied marked as a dead end;

            go back to node K_j; $w := w - w_{ij}$; $i := k$; go to $(\star)$

        end if

    else if $m_{ij} =$ free **then**

        {– a node is unexpectedly visited again – }

        $m_{ij} :=$ occupied {– it is a dead end – };

        drive back to K_j; $w := w - w_{ij}$; $i := j$: go to $(\star)$

    else if $m_{ij} =$ visited **then**

        {– draw back from a dead end – }

        $m_{ij} :=$ occupied

    else if there is an edge s_{ik} marked free **then**

        $m_{ik} :=$ visited; drive to node K_k;

        $w := w + w_{ik}$; $i := k$; go to $(\star)$

    else

        {– all edges up to s_{ir} are marked occupied – }

        $m_{ir} :=$ occupied; drive back to K_r;

        $w := w - w_{ir}$; $i := r$; go to $(\star)$;

    end if

  end if

end for

7.2 Local path planning

Depending on the task specification and the representation of the environment the local path planning defines how a vehicle can traverse a given environment. The following sections present different aspects in this field:

Local path planning can be done on different *representations of the environment,*

- Geometric maps: either with large free room available, or with narrow passages

- Raster maps: by flooding the raster or using quadtree representations in a one or two step planning

Then there are different *goals of planned paths*:

- Area covering paths on polygon maps driving along a wall

- Exploration of the environment: every part of the environment should be inspected at least once

Another more fundamental aspect is *obstacle avoidance* with movements to surround an obstacle using

- a wandering standpoint algorithm or

- a potential field method.

There are some basic abilities a vehicle should have:

- Driving an s-curve

- Forward and backward docking

- Path planning under geometric restrictions imposed by the vehicle itself

- Calculating velocity and turning rates

Finally a little bit beyond the scope of this chapter is looking for a stable equilibrium of a planned path: small deviations should not lead to completely different paths.

7.2.1 Path planning on geometric maps

The task of path planning on geometric maps is to define path points on a way from start to goal as paths which are suitable for a vehicle. As the dimensions of the vehicle and its kinematic restrictions begin to play a role, they must be taken into account.

Large free space

Let us assume that the free space is large compared to the dimensions of the vehicle. Let further the distance to an obstacle be not smaller than d_{min} and $d_{min} >>$ vehicle dimensions. Then planning a path around an obstacle is shown in figure 7.10.

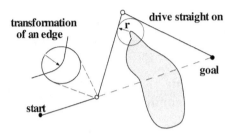

Figure 7.10 Path planning around an obstacle

At the most prominent part of the obstacle, a circle with $r = d_{min}$ is drawn. Go towards the goal as far as possible: a tangential line from there to the circle does not come closer to the obstacle than d_{min}. This is a first path point. Let the same be true for the tangential line from the goal. Then a second path point is the crossing point of both tangents. Between path points, the vehicle can drive straight on. Due to the fact that the vehicle has to change its direction at the path points, the "point" really is a curved path, dictated by the velocity of the vehicle and its kinematic restrictions as the inset in figure 7.10 shows. Under the assumption of large free space, this is no problem.

Driving around the obstacles is realized as shown in algorithm 7.3 and sketched in figure 7.11 which is an extension of figure 7.10.

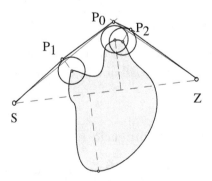

Figure 7.11 Iterative path planning around complex obstacles

Algorithm 7.3 Driving around an obstacle

check the line between start and end point $B = S$ and $E = Z$
($**$)
if the line cuts through an obstacle **then**
    look for points farthest away from the line
    take the side with the point of smallest distance to the line
    as shown in figure 7.11;
    Draw a circle with $r = d_{min}$ around this point;
    ($*$)
    Draw the tangents to this circle from B and E. They will meet at point
    $M = P_0$.
    check the lines $B - M$ and $M - E$ for obstacles going to ($**$);
else
    if if there is no obstacle on the way **then**
        look for the point of minimal distance to the line
        draw a circle with $r = d_{min}$ around this point
        if the circle cuts the line **then**
            go to ($*$)
        end if
    end if
end if

Finding path points according to Schweikard [SW95] Originally
Schweikard planned paths for a robot arm reaching around obstacles. This
may be extended for the movement of a vehicle around obstacles. Path points
need to be found which allow a path of width $2h$ to pass around all obstacle.
The algorithm runs as given in algorithm 7.4. Figure 7.12 shows the algorithm handling a convex obstacle, figure 7.13 a concave obstacle. Here the
distance update δ is larger than the circles radius d.

The algorithm finds a path from a starting point S to a goal G.

Figure 7.14 shows two obstacles with different possible paths running
from starting point S to goal G.

Grow obstacles according to Lozano-Perez [LP79] Let the vehicle
be round with its radius $r < d_{min}$ or it may be treated as a point if d_{min} is
very large compared to the vehicle dimensions. In both cases the obstacles
may be enlarged by d_{min}. Then the calculation looks for path points only
and any calculations of tangents and minimal distances may be skipped as
shown in figure 7.15

Algorithm 7.4 Path points according to [SW95]

$P := S; Q := G$

if the line $P \rightarrow Q$ cuts an obstacle **then**

　{as shown in figure 7.12}

　form a line through the cutting point at right angles

　if the line goes into free room **then**

　　place a path point P' at distance d there;

　　check $(P \rightarrow P')$ and $(P' \rightarrow Q)$ wether they cut an obstacle

　else

　　{the line cuts through the obstacle for a distance δ going into free room again}

　　place a path point P'' at distance $\delta + d$ into free room;

　　check $(P \rightarrow P'')$ and $(P'' \rightarrow Q)$ wether they cut an obstacle

　end if

else

　{$(P \rightarrow Q)$ does not cut an obstacle}

　check a path of width $2h$ between P and Q;

　if the side lines do not cut an obstacle **then**

　　$(P \rightarrow Q)$ is an established path;

　　check for the next part of the path;

　else

　　{one of the side lines of the path cuts through an obstacle}

　　form a line at right angles through the cutting point;

　　place a path point P' at a distance h into the free room;

　　check the paths $(P \rightarrow P'), (P' \rightarrow Q)$ if they cut an obstacle

　end if

end if

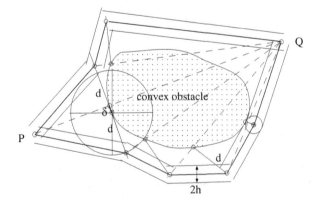

Figure 7.12 Path points around a convex obstacle

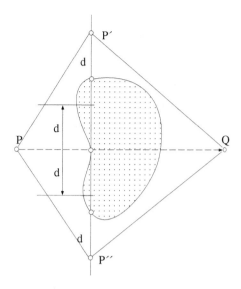

Figure 7.13 Planning path points around a concave obstacle

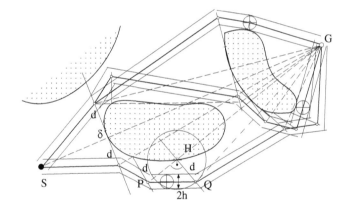

Figure 7.14 Determination of path points according to [SW95]

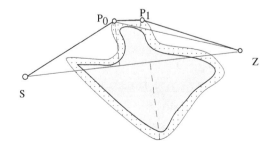

Figure 7.15 Grown obstacles

Narrow free room

In case of a narrow free room the vehicle has to move between nearby obstacles like a boat on a small river. The best thing to do is to move in the
middle of the free room. The distances between opposite shores are still large
compared to the vehicle dimensions. Let the map be given as a polygon map.
The result of path planning is a set of path points as shown in figure 7.16.
The search for path points follows this algorithm:

- look for convex obstacle points

- build the normal through this point towards the free space

- take into account only normals with foot points at the other shore

- the mid points of these normals are the path points wanted

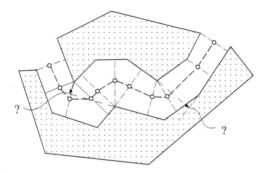

Figure 7.16   Narrow river path points

Finding a path point   Let (R,P) and (P,S) be polygon lines on one shore,
(P_1,P_2) be a polygon line at the other shore according to figure 7.17. Let the
angle $(S,P,R) = \gamma < 180°$ then P is a convex point. Let Q be the footpoint
of the normal on (P_1,P_2) through P. Then

$$\tan \varphi = \frac{y_2 - y_1}{x_2 - x_1}$$
$$d = (y_2 - y_1 + \tan \varphi \cdot (x - x_1)) \cos \varphi \qquad (7.1)$$
$$x_q = x + d \sin \varphi$$
$$y_q = y - d \cos \varphi$$

If $(x_1 \leq x_q \leq x_2) \wedge (y_1 \leq y_q \leq y_2)$ then there is a path point K with
$k_x = x + (x_q - x)/2$ and $k_y = y + (y_q - y)/2$, else look for a new line.

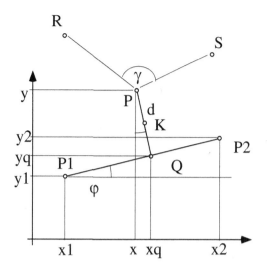

Figure 7.17 Finding a path point at a narrow

Mid point between smooth shores Let the map be given as continuous curves $f_1(x)$ and $f_2(x)$, e.g. splines through path points. The mid points between the shores may be found by literally pressing a balloon through the narrows: its mid point defines the curve with maximal distance to the shores as sketched in figure 7.18. Let P be a convex point on $f_1(x)$ defined by $(d^2 f_1(x))/(dx^2) < 0$. Form the normal at P: $y = mx+b$ with $m = -1/f_1'(x_p)$ and $b = y_p - mx_p$. Form the normal from a point Q from $f_2(x)$ cutting the normal on P at M. Vary Q until $r_p = |M,P| = r_q = |M,Q|$. Then M is the mid point of the touching circle and one of the points wanted. The algorithm 7.5 runs as shown in figure 7.19.

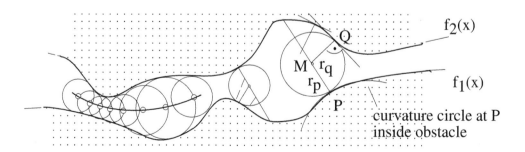

Figure 7.18 Path points as mid points of balloons

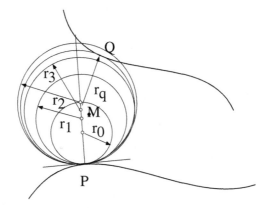

Figure 7.19 Finding the radius of the touching circle

Algorithm 7.5 Finding the radius of the balloon

$r := r_0; \Delta r = r_0;$
$(*)$
form a circle with r cutting P with its center on
line orthogonal to tangent in P;
if the circle cuts $f_2(x)$ **then**
    if the distance of the cutting points is $< \varepsilon$ **then**
        $r_q := r;$ M is a path point
    else
        $\Delta r := \Delta r/2;$ $r := r - \Delta r;$ go to $(*)$
    end if
else
    $r := r + \Delta r;$ go to $(*)$
end if

Elastic bands The elastic band method introduced by [QK93] is used to adapt a planned path online according to the robot motion and dynamic obstacles. The band consists of n bubbles b_i, $i = 1, \ldots, n$, representing the free space along the path where the robot can move. An example based on a gridmap representation of obstacles is shown in figure 7.20.

In the following, the main idea and algorithmic procedure are introduced according to [PS03]. Each bubble is defined by its center $\vec{c}_{bi}$, radius r_{bi} and obstacle masking distance D_{mi}. The bubbles are spread along the path to guarantee an overlapping according to the robot dimensions (see algorithm 7.6). Thus, the area covered by the bubbles mark the collision-free motion space of the robot.

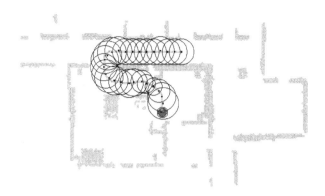

Figure 7.20 Elastic band along path representing free space between obstacles

$$L_i = \sum_{j=1}^{i} \| \vec{c}_{bj-1} - \vec{c}_{bj} \| \qquad (7.2)$$

The masking distance D_{mi} is related to the position L_i of the bubble along the path according to equation 7.2. It specifies the range around each bubble center in which obstacles are ignored (see equation 7.3).

$$D_{mi} = D_{m,\max} \cdot \begin{cases} 0 & \text{if } L_i \leq L_{\min} \\ 1 & \text{if } L_i \geq L_{\max} \\ \dfrac{L_i - L_{\min}}{L_{\max} - L_{\min}} & \text{otherwise} \end{cases} \qquad (7.3)$$

The parameters $L_{\min}$ and $L_{\max}$ represent the stretching limits of D_{mi} (accumulated path lengths). $D_{m,\max}$ is the maximum range at which obstacles are neglected. Suitable parameter values are listed in figure 7.21.

Obstacles are represented by the center points $\vec{p}_j = (x_{pj}, y_{pj})$ of the grid cells. Each bubble is related to a set of masked obstacles $\{\vec{p}_{m,ij}\}$ according to equation 7.4.

$$\{\vec{p}_{m,ij}\} = \{\vec{p}_j : \| \vec{c}_{bi} - \vec{p} \| > D_{mi}\} \qquad (7.4)$$

The obstacle $\vec{p}_i^*$ closest to b_i is the critical one (see equation 7.5).

$$\vec{p}_i^* = \arg(\min_{\vec{p} \in \{\vec{p}_{m,ij}\}} \| \vec{c}_{bi} - \vec{p} \|) \qquad (7.5)$$

It defines the bubble radius r_{bi} (space guaranteed to be free) according to equation 7.6.

$$r_{bi} = \min_{\vec{p} \in \{\vec{p}_{m,ij}\}} \| \vec{c}_{bi} - \vec{p} \| \qquad (7.6)$$

parameter	value
$L_{\min}$	2000 mm
$L_{\max}$	8000 mm
$D_{m,\max}$	0.0 mm
$r_{\lim}$	1600.0 mm
α_{int}	10.0
α_{ext}	10.0
ε	50.0 mm
to compare:	
robot radius	400.0 mm

Figure 7.21   Parameters and exemplary values used for elastic band algorithm

Based on these relationships, the bubbles are completely defined. The first step of the elastic band method is to translate a given path into a minimal set of bubbles covering the free space, corresponding to the robot dimensions. The next step is to adapt these bubbles in size and position while the robot is moving along the path. For that purpose, the first bubble follows the robot and the last one is stuck fixed at the goal. All other bubbles move iteratively due to the influence of two forces $\overrightarrow{f}_{int}$ and $\overrightarrow{f}_{ext}$ on their center $\overrightarrow{c}_{bi}$. This shift from time t to $t+1$ is given in equation 7.7.

$$
\begin{aligned}
\overrightarrow{c}_{bi,t+1} &= \overrightarrow{c}_{bi,t} + \Delta\overrightarrow{c}_{bi} && \text{with} \\
\Delta\overrightarrow{c}_{bi} &= \alpha_{tot,i} \cdot \left(\overrightarrow{f}_{int,i,i-1} + \overrightarrow{f}_{int,i,i+1} + \overrightarrow{f}_{ext,i}\right) && \text{and} \\
\alpha_{tot,i} &= \begin{cases} 1 & \text{if } r_{bi} > r_{\lim} \\ \dfrac{r_{bi}}{r_{\lim}} & \text{otherwise} \end{cases}
\end{aligned}
\tag{7.7}
$$

The internal force $\overrightarrow{f}_{int,ij}$ determines the strength of cohesion between adjacent bubbles i,j (see equation 7.8).

$$
\overrightarrow{f}_{int,ij} = \alpha_{int} \cdot \begin{cases} 0 & \text{if } \|\overrightarrow{c}_{bi} - \overrightarrow{c}_{bj}\| \le \varepsilon \\ \dfrac{\overrightarrow{c}_{bj} - \overrightarrow{c}_{bi}}{\|\overrightarrow{c}_{bi} - \overrightarrow{c}_{bj}\|} & \text{otherwise} \end{cases}
\tag{7.8}
$$

The external force $\overrightarrow{f}_{ext,i}$ represents a repulsion of bubble i from its critical obstacle $\overrightarrow{p}_i^*$ (see equation 7.9).

$$
\overrightarrow{f}_{ext,i} = \alpha_{ext} \cdot \begin{cases} 0 & \text{if } r_{bi} \le \varepsilon \text{ or } r_{bi} \ge r_{\lim} \\ \dfrac{r_{\lim} - r_{bi}}{r_{bi}}(\overrightarrow{c}_{bi} - \overrightarrow{p}_i^*) & \text{otherwise} \end{cases}
\tag{7.9}
$$

The parameter r_{lim} specifies the distance limit at which the elastic band is influenced by obstacles, α_{int} and α_{ext} are force weighting factors, $\alpha_{tot,i}$ creates a proportional relation between bubble size and agility and ε avoids division by zero. Figure 7.21 lists all parameters introduced so far and their values used during tests with mobile robot MARVIN (see figure 7.20).

Based on equations 7.7 to 7.9, the bubble centers are iteratively moved at each cyclic calculation step. Their respective radii are calculated according to equation 7.6, taking possible dynamic obstacles into account. Whenever the robot-fixed bubble is approximately coincident with its neighbor, it is deleted and replaced by this one. Consequently the band represents a dynamically smoothed version of the path around obstacles from the robot to the goal position. Its changing thickness marks narrow passages and free areas where the robot can move faster.

The whole procedure stops when the robot has reached the goal of the path (last bubble). Algorithm 7.6 summarizes the different calculation steps.

Algorithm 7.6 Elastic band algorithm

{**Given:**}
1. grid map with obstacles as center of occupied cells $\vec{p}_j = (x_{pj}, y_{pj})$
2. path represented by center of visited grid cells

{– **preparation** – }
calculate minimal set of n bubbles b_i with center $\vec{c}_{bi}$, radius r_{bi}
- $\vec{c}_{b1}$ = robot pose
- $\vec{c}_{bn}$ = goal pose
- $\| \vec{c}_{bi+1} - \vec{c}_{bi} \|$ according to robot dimension
- r_{bi} according to equation 7.6

{– **online update** (robot moving along path)– }
while robot is not at goal **do**
    update grid map (dynamic obstacles)
    remove b_1 if $\| b_2 - b_1 \|$ too small
    for every remaining bubble **do**
        determine critical obstacle $\vec{p}_i^*$
        calculate r_{bi} according to equation 7.6
        calculate external and internal forces
        apply shift to bubble center (cf. equation 7.7)
    end for
end while

One remaining problem is what to do if some bubbles shrink below robot dimensions. In this case, there is not enough free space for the robot to follow the path. When this situation is stable for a certain amount of time, a replanning has to be triggered to take new obstacles into account. Of course, the planning should not be started immediately when it gets invalid as obstacles might vanish before the robot is getting close to them (e.g. a person crossing the path). The new plan is then given as input for a new run of the elastic band algorithm.

Generalized cones   Following **Brooks** [Bro82] in a map built from polygons a path with maximum distance to obstacles may be found using generalized cones as shown in figure 7.22. Look into the free space; find half angle lines between opposite shores, they define points of equal distance to either shore. Let P be a convex point. The normal to the other shore hitting P defines a narrow in the free room. Let the angle between the counter shore and the line through P be $< 90°$ as shown in the left part of the drawing. Draw a parallel line to the counter shore through P. The parallel line at half the distance of the narrow hits the half angle line at a way point. In the right part of the drawing, the situation with an angle $> 90°$ is shown: take the other half angle line: it cuts the normal to the counter shore through P forming a way point for this type of narrow in the path.

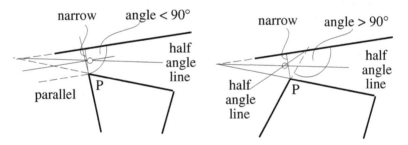

Figure 7.22   Generalized cones

Half angle line   Following figure 7.23 let the shores be defined by lines (P_1, Q_1) and (P_2, Q_2).
    A line through P_1 an Q_1 is given by

$$y = \frac{y_{q1} - y_{p1}}{x_{q1} - x_{p1}}(x - x_{p1}) + y_{p1} \tag{7.10}$$

$$y = \alpha_1(x - x_{p1}) + y_{p1} \tag{7.11}$$

A line through P_2 is given by

$$y = \frac{y_{q2} - y_{p2}}{x_{q2} - x_{p2}}(x - x_{p2}) + y_{p2} \tag{7.12}$$

$$y = \alpha_2(x - x_{p2}) + y_{p2} \tag{7.13}$$

The cutting point $S = (x_s, y_s)$ is given by
$(\tan\alpha_1 - \tan\alpha_2)x_s = \tan\alpha_1 x_{p1} - \tan\alpha_2 x_{p2} + y_{p2} - y_{p1}$

$$x_s = \frac{\tan\alpha_1 x_{p1} - \tan\alpha_2 x_{p2} + y_{p2} - y_{p1}}{\tan\alpha_1 - \tan\alpha_2} \tag{7.14}$$

$$y_s = \tan\alpha_1(x_s - x_{p1}) + y_{p1} \tag{7.15}$$

The half angle line then is

$$\boxed{y = \tan(\alpha_2 + \alpha_1)/2 \cdot (x - x_s) + y_s} \tag{7.16}$$

and can be used to calculate the waypoint P lying at the intersection of two half angle lines.

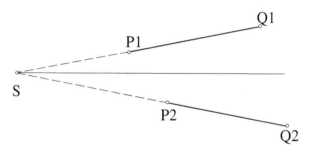

Figure 7.23 Half angle line

7.2.2   Navigation on a raster map

A map is given in form of a binary raster with obstacles and free space only. The task is to find the shortest way from a starting point to a goal. At first the raster is flooded beginning with the starting point as shown in figure 7.24. A wave runs around obstacles and fills the whole obstacle free room. From the goal point go downwards until the starting point is reached again forming a list of passed raster points. Then go again through this list: it is a shortest path from start to goal. Figure 7.25 shows the first three steps in flooding a raster in a yet empty map.

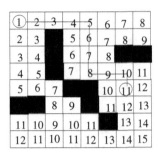

Figure 7.24 Flooding a raster map

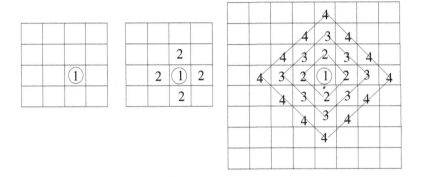

Figure 7.25 Flooding an empty raster

Using the neighbor points as shown in figure 7.26, an algorithm to flood a raster runs as given in algorithm 7.7. If a raster point got a number, it keeps that number. In order not to go through all the raster points many times the algorithm uses two stacks S_1 and S_2 which are filled and emptied alternately. Cells denoting free room are $z = 0$, obstacle cells $z = -1$. In the beginning, both stacks are empty.

```
      N
  W   z   O
      S
```

Figure 7.26 Neighbor points of a raster point z

As an example, solving a maze in a raster map is shown in figure 7.27 by flooding the raster. The assumption here is that the passages are broad compared to the thickness of walls as shown in the figure.

Algorithm 7.7 Flooding a raster with thick walls

Initialization: $i := 1$; $j := 0$;
starting cell $z := 1$; push S_i;
repeat
   $i := (i+1) \mod 2$; $j := (j+1) \mod 2$
  repeat
    pop stack S_i; check surrounding of the cell taken out of stack:
    for all cells with $z = 0$ **do**
      if $\min_{z>0}(O,S,W,N) > 0$ **then**
        $z := \min_{z>0}(O,S,W,N) + 1$; push to stack S_j;
      else
        do nothing;
      end if
    end for
  until stack S_i is empty
until stack S_j is empty

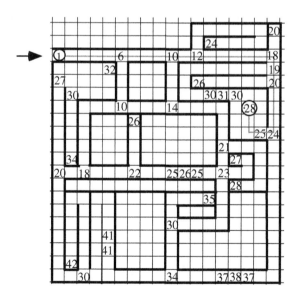

Figure 7.27 Solving a maze by flooding a raster map

7.2.3 Quadtree based path planning

One step planning A map is given as a quadtree with white and black nodes. Then path planning with an A*-algorithm can be undertaken with

- expansion of a node: look for neighboring nodes with free room

- expansion of a node in the next step

- expansion of vertical or horizontal obstacle-free neighboring nodes

- expand the node with the minimal value of the weight function

The weight function is

$$f(C) = g(P) + d(P,C) + \alpha \cdot (O_{\max} - O(C)) + h(C) \qquad (7.17)$$

with

C: obstacle-free node

P: predecessor of C on this path

$g(P)$: costs of the path from starting point S to P

$d(P,C)$: actual distance from P to C

$O(C)$: distance from C to the next obstacle

$O_{\max}$: maximal distance from any C to the next obstacle

α: constant, describing the minimally allowed distance value to obstacles

$h(C)$: optimistically guessed distance between C and the goal Z

The result is a list of obstacle-free nodes of different sizes. The merits of this one step planning are

- efficient calculation of the path

- α gives the minimum distance to obstacles

- no large overhead in the quadtree representation

Two step path planning   Let a map be given as a quadtree representation with white, gray and black nodes.

- the path is planned on as high a level as possible

- in a second step the gray nodes are expanded

- gray nodes are expanded only to the extent that is absolutely necessary

- smallest squares have vehicle dimension

    The merits of a two step planning are:

- the algorithm is cost effective if many small obstacles allow only a few large white nodes

- unknown regions may be treated efficiently as gray nodes with large costs

7.2.4   Area covering paths

Let a raster map be given. The vehicle has the task to clean or paint (with a brush) all free raster points. Within a raster point, the vehicle can turn or change its direction. The raw version of a cleaning algorithm runs as given in algorithm 7.8.

Algorithm 7.8 Area covering algorithm

$(*)$
try to drive straight paths
check parallel raster points
mark beginning and end points of paths on a stack
mark points already driven through
if the path ends **then**
    {there is either an obstacle or a marked raster point hit}
    pop the stack
else if the stack is not empty **then**
    the mark of the beginning of a path not yet driven is taken from the stack
    drive to that point; go on at $(*)$
else
    the area is filled
end if

Figure 7.28 shows a typical situation in an area covering algorithm. The open circles denote starting points to be laid down on the stack with the direction of where to go later. If there is no free room left at the moment, the robot will drive through a region already cleaned to a next starting point. An example of covering an area using a raster map is given in figure 7.29.

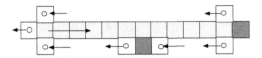

Figure 7.28 Typical situation during area covering using a grid map

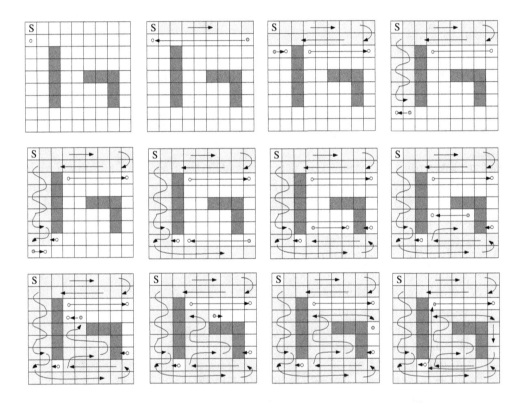

Figure 7.29 Example of area covering using a raster map

Area covering using a polygon map A typical application for area covering navigation is the task to clean or paint a room by using a brush of a suitable diameter. For such an application neighboring paths should sufficiently overlap. Assuming that a polygon map of a room is given, algorithm 7.9 runs as shown in figure 7.30.

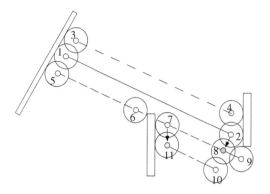

Figure 7.30   Determination of overlaying paths in a polygon map for cleaning applications

---

Algorithm 7.9 Area covering algorithm in a polygon map

---

initialization: find the main direction in the map

begin at a wall to plan paths in that direction (1)

repeat

    place path end points to the left and right in free room parallel to that path on the stack (5), (3);

    if there is an obstacle to the right or the left **then**

        place beginning points to the right or the left of the planned path on the stack (6), (4);

    else if there is free room again **then**

        place path end point on the stack (7);

    else if the path hits a wall **then**

        place path beginning points on the stack (2);

    else

        pop path beginning- or end-point from stack; plan a path to there through already cleaned area; follow the path given by top of stack (8);

    end if

until stack is empty

To produce such an area covering path, only four basic abilities of the vehicle are required as shown by **Hofner** [Hof97] sketched in figure 7.31.

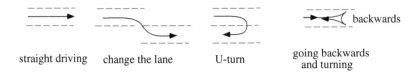

straight driving   change the lane   U-turn   going backwards and turning

Figure 7.31 Elementary abilities to fill an area

Area covering by following a wall In a continuous map area covering paths may be constructed according to algorithm 7.10 as sketched in figure 7.32.

Algorithm 7.10 Area covering by following a wall

---

start at a wall, follow that wall and mark the path driven (i.e by painting the floor);
repeat
    if the wall changes into into a painted path **then**
        follow that path (spiralling inwards);
        if an obstacle is hit **then**
            follow that new wall – It might be necessary to change direction -.
            put this corner point on a stack;
        else
            {there is no room left} Put this corner point on a stack too
            take a corner point from the stack and drive to this point through
            an already painted region
        end if
    end if
until there is no corner point left
the floor is painted;

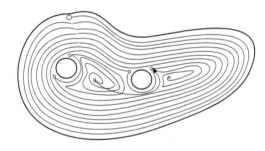

Figure 7.32 Area covering path by following a wall

7.2.5   Exploration

The task is to find a path covering all parts of a map. The vehicle has dimensions r, a sensor reaching R out into the environment and typical distances d in its vicinity. Let $r \ll d < R$. The idea is to find points of interest (POI) in the vicinity of the vehicle to drive to for exploring the environment. Seen from the vehicle, a typical sensor situation is shown in figure 7.33. The exploration is performed as described in algorithm 7.11. An example exploration is given in figure 7.34.

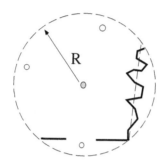

Figure 7.33 A typical sensor situation

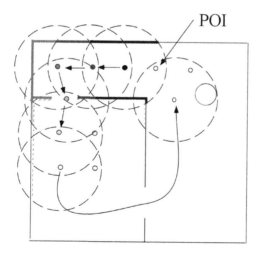

Figure 7.34 Driving in a circle

Algorithm 7.11 Exploration

$(*)$

check the map from the momentary point of view

if there are through ways between obstacles (width $\geq 3r$; beyond is free room up to R) **then**

    place points of interest (POI) at a distance $\approx 4/5R$ from the vehicle in the middle of through ways; mark them as "free"; pack them on a stack; {– Beyond the POI there should be room left for the vehicle to turn back – }

end if

if there are no obstacles around the vehicle **then**

    place POIs in the free room at an angle of 90° to other POIs in the free room

    at a distance of $2/3R$ keeping away $\approx 3/2r$ from obstacle; {– the distance is taken again to give the vehicle room for maneuvers beyond, more than in through ways between obstacles – }

    mark them as "free"; pack them on a stack;

end if

if within d there is a POI marked as already visited **then**

    a loop has been closed;

end if

$(**)$

if the stack is not empty **then**

    pop POI from stack;

    if the POI is marked as "free" **then**

        mark it as "visited"; drive through explored regions to this POI;

        go on at $(*)$;

    else

        {a loop has been detected}

        go on at $(**)$

    end if

else

    {the map is explored}

end if

7.2.6   Obstacle avoidance

One of the basic abilities of a vehicle should be driving around obstacles even if the available map does not them all. At a given velocity v obstacles at a distance $d > r_0$ may be circumvented: Any obstacle nearer than r_0 in the path of the vehicle leads to a collision if no emergency break is commanded. Figure 7.35 shows an example of obstacle avoidance.

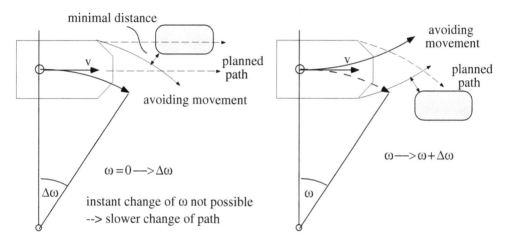

Figure 7.35 Maneuvers for avoiding an obstacle

In the left part of the drawing the vehicle runs straight forward. Then $\omega = 0$. Insert a $\Delta\omega$ so that a minimum distance to the obstacle is kept. The right part of the figure shows a vehicle just turning at $\omega \neq 0$. Insert a $\Delta\omega$ so that the vehicle passes the obstacle at a safe distance. Figure 7.36 shows the complete turn around that obstacle. The same figure shows the paths on a larger scale, too. At the end of the deviating movement the vehicle follows the originally planned path again.

Wandering standpoint algorithm, WSA   Often a goal may be seen from far away but there are obstacles on the way to the goal. Given a maximal sensor distance r_s and a minimal distance $r_0 << r_s$ where a deviation from a straight path has to be commanded, a wandering standpoint algorithm (see algorithm 7.12) is sketched in figure 7.37.

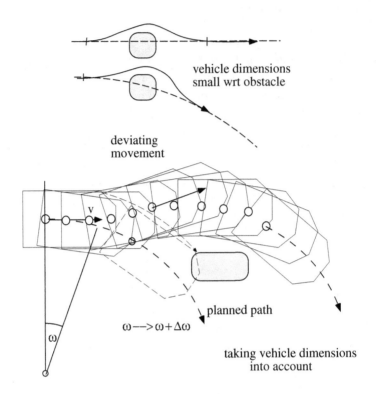

Figure 7.36 Driving around an obstacle

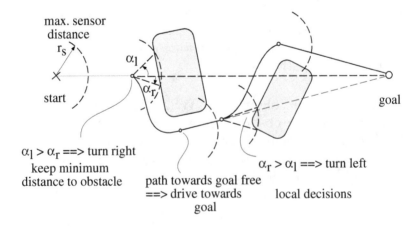

Figure 7.37 Avoiding obstacles using the WSA

Algorithm 7.12 Wandering standpoint algorithm

$(*)$

run towards the goal until the goal is reached

if the sensor detects an obstacle at a distance $d > r_0$ **then**

   drive towards the obstacle till $r_s < r < r_0$;

   measure the angles α_l and α_r to the obstacle;

   turn towards the smaller angle and keep the distance to the obstacle constant;

else

   go to $(*)$; {the direction to the obstacle is free again}

end if

Apart from pathological situations, this strategy will find a way to the goal. There are two possible cases of failure:

- A movable obstacle may push the vehicle away from the goal like a defender pushes away an attacking player from the goal.

- A pathological form of the obstacle lets the algorithm fail as figure 7.38 shows:

  The form of the border line lets the vehicle drive until the goal can be seen again, but the distance to the goal remains constant, though the vehicle went straight towards the goal for a while.

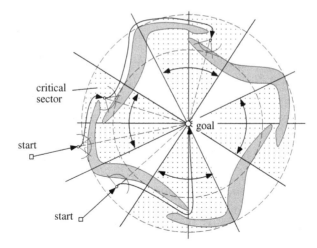

Figure 7.38 A pathological case for WSA

In order to cope with this case, the vehicle should know its overall turning: having turned 360° either the distance to the goal has become smaller then the vehicle is spiralling inwards according to figure 7.39, or the distance stood constant: then for a moment prefer to turn to the larger angle: if there is a way towards the goal this altered strategy will find it.

Figure 7.39   An obstacle in form of a spiral

7.2.7   Potential field method

Forces on a Vehicle   A sector map with obstacles is given. At any point (x,y) the influence of an obstacle is modeled by a retarding force felt by the vehicle, driving the vehicle away from the obstacle as shown in figure 7.40. The force added by a sector with an obstacle at distance d_i in sector φ_i is $F_i = \cos\varphi_i \frac{1}{d_i}$. At the same time a driving force F_Z is felt, drawing the vehicle towards the goal.

The total force is $F_g(x,y) = F_Z(x,y) + \sum_{i=1}^{n} F_h^i(x,y)$ with n obstacles in the near vicinity of the vehicle. At any moment the vehicle is running in the direction of the total force. F_g is a potential field: it may be described by the gradient of a scalar field: $F_g(x,y) = -\mathrm{grad}\phi(x,y)$. The gradient describes the direction of steepest incline of a scalar field: $\mathrm{grad} = (\partial\phi/\partial x, \partial\phi/\partial y)$. As long as the driving force towards the goal is large enough, the vehicle will follow this force and even surmount a ridge in between.

Polar histogram   Instead of calculating the forces, compile a polar histogram of the obstacle density according to figure 7.41.

Look for a minimum in the histogram: If the obstacle density is zero there, then this is the intended control angle φ^*. Choose $\omega(t)$ such that $\varphi^*(t) = \int_0^t \omega(\tau)d\tau$. This task may be solved for a given change in direction $\Delta\varphi$ using a special form of $\omega(t)$: let $\omega(t) = \omega_0 \cdot (1 - \cos(2\pi \cdot t/T))$.

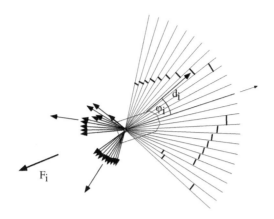

Figure 7.40 Potential forces on a vehicle resulting from obstacles and goal point

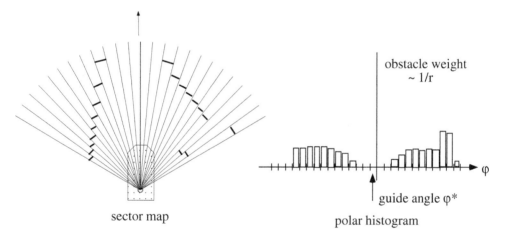

sector map   polar histogram

Figure 7.41 A polar histogram based on a sector map

Then $\Delta\varphi = \int_0^T \omega(\tau)d\tau = \omega_0 \cdot T$. The time T is restricted by the maximal angular acceleration

$$dw/dt = \omega_0 \cdot 2\pi/t \cdot \sin(2\pi/T \cdot t); \tag{7.18}$$
$$(dw/dt)_{\mathrm{max}} = \omega_0 \cdot 2\pi/T \tag{7.19}$$

Now the angular velocity ω_0 and the time T may be calculated from

$$\omega_0^2 = \Delta\varphi \cdot (dw/dt)_{\mathrm{max}}/2\pi \tag{7.20}$$
$$T^2 = 2\pi \cdot \Delta\varphi/(dw/dt)_{\mathrm{max}} \tag{7.21}$$

To change direction by $\Delta\varphi$ without changing velocity takes time T and the vehicle will have run through a distance $s = v \cdot T$ before the new direction is reached.

To ease the calculations, the functions $\omega(t)$ and $d\omega/dt$ may be calculated beforehand as shown in figure 7.42. It is a list of 1024 values spanning the time from $0 \longrightarrow T$, read off every δt seconds interpolating in the list and given as commands to the vehicle.

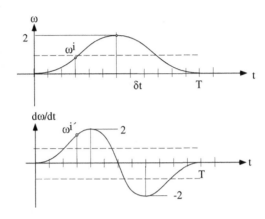

Figure 7.42 Function to drive smoothly through a curve

7.2.8 Basic abilities

Aside from avoiding obstacles, there are some basic abilities an autonomous vehicle should have:

change direction by $\Delta\varphi$ As shown before, this is done commanding a turning by $\omega(t)$ for a time T until the integral of ω is the wanted angle $\Delta\varphi$

drive through an s-curve with Driving through an s-curve with an an inclination angle φ_0 and a deviation d as shown in figure 7.43, is realized by driving on a circle with radius $r = v \cdot \omega_0$ until the angle driven on the circle is φ_0, then switching to a circle with the same radius r but driving to the other side as shown in figure 7.44. This form of movement is a simplification of reality: it is not possible to change $\omega(t)$ instantly.

Once driving through an s-curve is possible, an application like a docking maneuver as shown in figure 7.45 is possible.

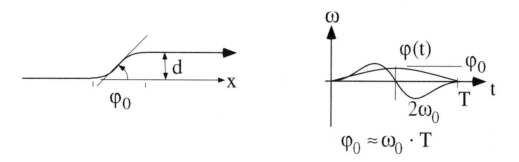

Figure 7.43 Driving an S-curve

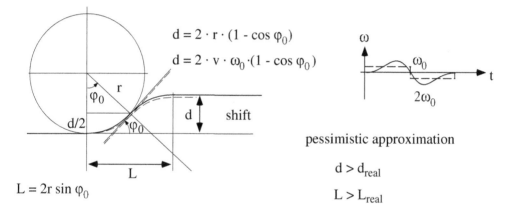

Figure 7.44 S-curve by two circles

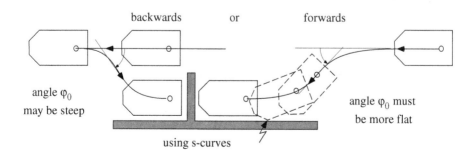

Figure 7.45 Forward and backward docking maneuver

Avoiding a Wall   Running straight towards a wall a deviating maneuver must be commanded: a turn of $90°$, then to run parallel to the wall according to figure 7.46. This picture is a simplification: the angular velocity cannot change from $0 \rightarrow \omega_0$ instantly. With $v= 1\,\text{m/s}$ and $\omega_0 = 45°/\text{s}$ and a commanded $\Delta\varphi = 90°$, the vehicle will still drive for $T = \Delta\varphi/\omega_0 = 2\,\text{s}$ and go for $s = 2\,\text{m}$ until the new path has been reached. During this time the vehicle came nearer to the wall by the radius of the circle driven: $r = 2/\pi \cdot v \cdot 90°/\omega_0\,\text{m} = 4/\pi\,\text{m} = 1.275\,\text{m}$. This plus a safety margin is the nearest distance to the wall where the deviating maneuver has to be commanded to avoid an emergency break.

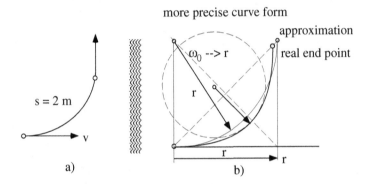

Figure 7.46   Driving towards an obstacle; (a) shows the general behavior, (b) the deviating maneuver in more detail as the vehicle cannot drive a curvature unsteady path.

Driving Through a Curve   Ideally, a vehicle should be able to turn on a spot: driving straight to a point P_0 and turning in a new direction. In fact the possible curve will look like figure 7.47 because of kinematic restrictions imposed by physics

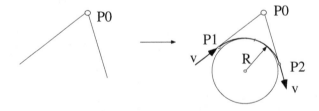

Figure 7.47   Kinematic restrictions driving through a curve

- the velocity v of the vehicle determines the radius R of the minimal curvature circle. Driving through a curve, side forces occur at the vehicle, limited by the allowed resistance against sliding. The side force $F_{side} \propto \omega^2$ and $\omega = v/R$

- going from a straight line into a curve needs a steady curvature behavior as shown in figure 7.48. To calculate the necessary steering function, go into a coordinate system with its x-axis through P_1 and P_2 and its y-axis through P_0 then

$$y'(x_0) = -\tan\alpha \qquad\qquad y'(-x_0) = \tan\alpha \qquad\qquad y'(0) = 0$$
$$y''(x_0) = 0 \qquad\qquad\qquad y''(-x_0) = 0$$

The curvature is

$$K(x) = \frac{y''(x)}{[1 + y'^2(x)]^{2/3}} \tag{7.22}$$

$K(0) = y''(0)$ with $|K(0)| = 1/R$

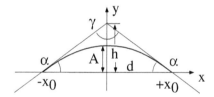

Figure 7.48 Steady curvature path

Two possible solutions to a curvature steady path will be given

- (1) a trigonometric curve

$$\boxed{y(x) = A \cdot \cos(\pi/2 \cdot x/x_0)} \tag{7.23}$$

$$y'(x) = -A \cdot \frac{\pi}{2x_0} \cdot \sin\frac{\pi x}{2x_0}$$
$$y''(x) = -A \cdot \left(\frac{\pi}{2x_0}\right)^2 \cos\frac{\pi x}{2x_0}$$

$$
\begin{aligned}
y(x_0) &= y(-x_0) = 0 \\
y'(x_0) &= -A \cdot \frac{\pi}{2x_0} = -\tan\alpha \qquad \Longrightarrow A = 2x_0/\pi \cdot \tan\alpha \\
y''(x_0) &= -A\left(\frac{\pi}{2x_0}\right)^2 \qquad \Longrightarrow R = \frac{4x_0^2}{\pi^2 A} = \frac{2x_0}{\pi\tan\alpha}
\end{aligned}
$$

$$
\boxed{x_0 = \pi/2R\tan\alpha \qquad A = R\tan^2\alpha} \tag{7.24}
$$

– path planning gives α

– kinematic restrictions give R

$\Longrightarrow x_0 = \pi/2 \cdot R\tan\alpha$ and $h = x_0 \cdot \tan\alpha$
The deviation of h from A is
$h - A = x_0 \cdot \tan\alpha(1 - 2/\pi) = R\tan^2\alpha(\pi/2 - 1)$

- (2) a polynom of degree 4

$$
\boxed{y(x) = ax^4 + bx^3 + cx^2 + dx + e} \tag{7.25}
$$

$$
y'(x) = 4ax^3 + 3bx^2 + 2cx + d \qquad\qquad y''(x) = 12ax^2 + 6bx + 2c
$$

$$
\begin{aligned}
y(x_0) = y(-x_0) = 0 && \Longrightarrow b = d = 0 \\
y'(x_0) = 0 && \Longrightarrow ax_0^4 + cx_0^2 + e = 0 \\
y(0) = A && \Longrightarrow e = A \\
y'(-x_0) = -4ax_0^3 - 2cx_0 && = \tan\alpha \\
y''(0) = 2c = -1/R && \Longrightarrow c = -1/(2R) \\
y''(x_0) = 12ax_0^2 + 2c = 0 && \Longrightarrow a = -c/(6x_0^2) = \frac{1}{12Rx_0^2} \\
y'(x_0) = -4ax_0^3 - 2cx_0 = \tan\alpha && \Longrightarrow x_0\left(\frac{1}{R} - \frac{1}{3R}\right) = \tan\alpha \\
\Longrightarrow x_0 = 3R/2 \cdot \tan\alpha &&
\end{aligned}
$$

$$
\begin{aligned}
\Longrightarrow (3R/2 \cdot \tan\alpha)2/(12R) - 1/(2R)(3R/2 \cdot \tan\alpha)^2 + e = 0 \\
e = 9/4 \cdot R^2 \tan^2\alpha(1/2R - 1/12R)
\end{aligned}
$$

$$e = 15/16 \cdot R \tan^2 \alpha \qquad\qquad x_0 = 3R/2 \cdot \tan \alpha \qquad\qquad p = x_0/\cos \alpha$$

In the end the polynom of degree 4 is

$$y = -\frac{1}{27R^3 \tan^2 \alpha} x^4 + \frac{1}{2R} x^2 + 15/16r \tan^2 \alpha \qquad (7.26)$$

Figure 7.49 shows a rather narrow curve and the relevant distances.

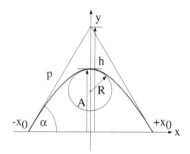

Figure 7.49   A rather narrow curve

For the curve of degree 4 the distances $h - A$ is

$$h - A = 3R/2 \cdot \tan^2 \alpha - 15/16 \cdot R \cdot \tan^2 \alpha \qquad (7.27)$$

$$h - A = 9/16 \cdot R \cdot \tan^2 \alpha \qquad\qquad p = \frac{3R \tan \alpha}{2 \cos \alpha} \quad (7.28)$$

For a trigonometric curve the same distances are

$$h - A = (\pi/2 - 1) \cdot R \cdot \tan \alpha \qquad\qquad p = \frac{\pi R \tan \alpha}{2 \cos \alpha} \quad (7.29)$$

The difference is $9/16 = 0.56$ and 3 vs $(\pi/2 - 1) = 0.57$ and π

Driving a path of given length   Following figure 7.49 the path length from point $(-x_0,0)$ to $(0,y_0)$ is given by the integral

$$s = \int_{-x_0}^{0} (1 + y'^2)^{1/2} dx$$

This integral is generally not solvable in a closed form. So in order to calculate s, numerical methods have to be used. For a vehicle it is important to find control functions $w(t)$ in real time, to drive through the wanted path at constant velocity. A set $w_i = w_i(t)$ is to be calculated at times $t_i = t_0 + i\Delta t$. Provided the curvature $K(x)$ is known, then $w(x) = v \cdot (x)$. The curvature is

$$K(x) = \frac{y''(x)}{(1 + y'^2)^{3/2}} \qquad (7.30)$$

To calculate w_i the curvature $K_i(x)$ is to be calculated at points x_i. They are a distance Δs apart with $\Delta S = v\Delta t = (1 + y''(x))^{1/2} \cdot \Delta x$ according to figure 7.50.

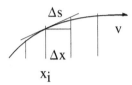

Figure 7.50 Δx and Δs for curvature calculation

In effect

$$x_{i+1} = x_i + \frac{v\Delta t}{(1 + y'^2)1/2} \qquad (7.31)$$

The integration of the curve length follows numerical methods to find the length s of the trigonometric function.

- Given a constant velocity v the time needed is $T = s/v$.

- Slow down the vehicle from v_1 so that at P_1 the velocity is v. Set $t = t_0$.

- During the time T produce a linear growing $w(t) = a(t - t_0)$ with $a = v/R \cdot T$. At $t = t_0 + T$, the angular velocity is $\omega = v/R$

- Slow down $w(t)$ for a time T from v/R to zero; then the vehicle is at P_2

- accelerate again to v_2

The time development of $w(t)$ is

$$w(t) = \begin{cases} \frac{v}{R \cdot T}(t - t_0) & t_0 \le t \le t_0 + T \\ v/R - v\frac{(t - t_0 - T)}{R \cdot T} & t_0 + T \le t \le t_0 + 2T \end{cases}$$

$$\varphi(t) = \int a(t - t_0)dt = (1/2at^2 - at_0 t) \qquad (7.32)$$

$$x(t) = a \int v \cos(1/2t^2 - t_0)dt \qquad (7.33)$$

$$y(t) = a \int v \sin(1/2t^2 - t_0)dt \qquad (7.34)$$

$x(t)$ and $y(t)$ describe a clothoid not readily integrable. The approximation in form of a cosine function for $y(t)$ is easier to handle, but there is an error involved. To cope with this, a correction is made as shown in figure 7.51.

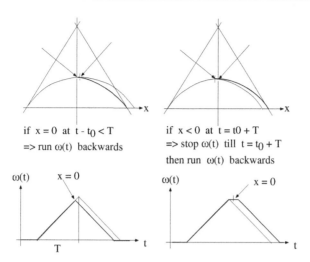

Figure 7.51 Method to correct the driving through a curve

Path planning under geometric restrictions   The restrictions are imposed on path planning by the vehicle geometry. If the vehicle is round, then
enlarging the obstacles by the radius r plus a safety margin δ will do. If the
vehicle is rectangular as shown in figure 7.52, with a length L, broadness B,
and a safety margin δ as well as a kinematic center which is not in the vehicle
center, then obstacle avoidance is performed using the following rules:

- When driving straight forward, just check for obstacles within a path-
  way of $B + \delta$.

- When driving on a curve $y = f(x)$, check for obstacles on a path (x^*, y^*)
  defined by the side movements of the vehicle according to figure 7.53.

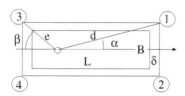

Figure 7.52 Parameter of a rectangular vehicle

Corner point 1 runs on a curve with $\varphi = \arctan f'(x)$ and $x_1^* = x + d\cos(\alpha + \varphi)$ and $y_1^* = y + d\sin(\alpha + \varphi)$.
Corner point 4 runs on a curve $x_4^* = x - e\cos(90° - \beta - \varphi)$ and $y_4^* = y - e\sin(90° - \beta - \varphi)$.

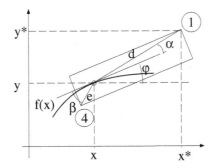

Figure 7.53 Obstacle checking based on vehicle parameters when driving through a curve

Planning a curved path for a long vehicle may need an appreciable free room to the side as shown in figure 7.54

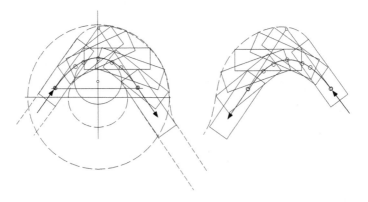

Figure 7.54 Path characteristics of a long vehicle

Driving through a door This is another basic ability a vehicle should have in an indoor environment. Finding a suitable passage could be taught to the vehicle by building a growing neural gas net. The nodes of the net are sensor situations, the position and orientation, and actions (v, ω) taken at a the moment. Edges of the net connect different sensor situations following each other by driving through a door. From different starting positions, the vehicle is commanded through a door. After a while the passages should be learned: in replay to a given sensor situation and position the vehicle should find the most similar node in the net. Interpolating the actions to the next action to be taken, the vehicle will find its way through the door as shown in figure 7.55.

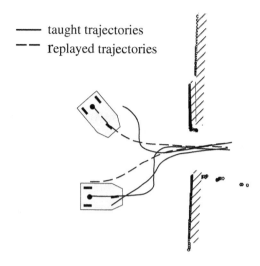

Figure 7.55 Passing through a door

Following a wall Using sector maps only, a vehicle may find its way between obstacles using polar histograms to calculate the steering angles as shown in figure 7.56.

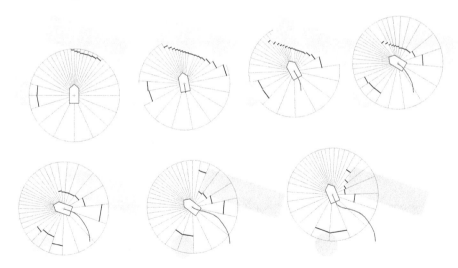

Figure 7.56 Wall following between obstacles

The vehicle tries to find a path between obstacles using polar histograms and tries to stay next to a wall,following a wall once detected.

7.3   Path control

Let a reference path z_r be given, with a reference point (x_r, y_r) at a time
t running at a reference velocity. The vehicle will find itself at a time t at
a position and orientation $(x, y, \varphi) = z(t)$ not on the reference path, according to figure 7.57. The task of path control is to find steering functions
$v^*(t)$ and $\omega^*(t)$ with which the vehicle goes from an arbitrary starting point
$(z(t_0), z'(t_0))$ into the reference path after a while: $\lim_{t \to \infty} z(t) = z_r(t)$.

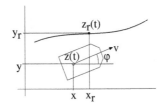

Figure 7.57   Vehicle, reference path, and reference point during path control

Let the velocities

$$x' = v \cos \varphi \qquad x(t_0) = x_0 \qquad\qquad \varphi' = \omega \qquad\qquad \varphi(t_0) = \varphi_0$$
$$y' = v \sin \varphi \qquad y(t_0) = y_0$$

and the accelerations

$$x'' = v' \cos \varphi - v\omega \sin \varphi \qquad\qquad\qquad x'(t-0) = x_0$$
$$y'' = v' \sin \varphi + v\omega \cos \varphi \qquad\qquad\qquad y'(t_0) = y_0$$

be measured. The error vector is $z_e = (x_e, y_e)^T = (x_r - x, y_r - y)^T$ and its
velocity v

$$x'_e = x'_r - v \cos \varphi \qquad\qquad x_e(t_0) = x_{e0} \qquad\qquad = x_{r0} - x_0$$
$$y'_e = y'_r - v \sin \varphi \qquad\qquad y_e(t_0) = y_{e0} \qquad\qquad = y_{r0} - y_0$$

and acceleration

$$x'' = x_r'' - (v \cos \varphi - v \cdot \omega \sin \varphi) \qquad\qquad x'_e(t_0) = x'_{e0}$$
$$y'' = y_r'' - (v \sin \varphi + v \cdot \omega \cos \varphi) \qquad\qquad y'_e(t_0) = y'_{e0}$$

Calculating $v(t)$ and $\omega(t)$ If $x''(t)$ and $y''(t)$ are given, $v(t)$ and $\omega(t)$ can be calculated. From $x'' = v' \cos \varphi - v \cdot \omega \sin \varphi$ and $y'' = v' \sin \varphi + v \cdot \omega \cos \varphi$

$$\Longrightarrow (y'' \cos \varphi - x'' \sin \varphi) = v \cdot \omega \tag{7.35}$$

$$\Longrightarrow (x'' \cos \varphi + y'' \sin \varphi) = v'(t) \tag{7.36}$$

Integrating $v'(t)$ the velocity $v(t)$ can be calculated and from $v(t)$ also $\omega(t)$

$$\Longrightarrow v(t) = v(t_0) + \int_{t_0}^{t} (x''(t) \cos \varphi(t) + y''(t) \sin \varphi(t)) dt \tag{7.37}$$

$$\Longrightarrow \omega(t) = \frac{y''(t) \cos \varphi(t) - x''(t) \sin \varphi(t)}{v(t)} \tag{7.38}$$

Looking for a stable equilibrium $(z_e, z_e') = 0$ If the vehicle stays on the reference path and follows the reference point, the error in position and velocity is and stays zero. In terms of control theory this is equivalent to the existence of a **Ljapunov function** $V(z_e, z_e')$ characterized by

- $V(z_e, z_e') = 0$ for $[z_e, z_e'] = 0$

- $V(z_e, z_e') > 0$ for $[z_e, z_e'] \neq 0$

- $V'(z_e, z_e') \leq 0$ for all $[z_e, z_e']$

If such a function exists, then (z_e, z_e') is asymptotically stable: for each starting point (z_0, z_0') and each reference path s with (x_r, y_r, x_r', y_r'), the reference point can be reached; $\lim_{t \to \infty} z(t) = z_r(t)$. Moreover, an additional functional form can be minimized: the weighted sum of the square error vector and its time derivative; this gives a smooth curve.

$$J(v(t), \omega(t), t) = \int_{t_0}^{t} \left(z_e^T z_e + z_e'^T R_1 z_e' + z_e'^T R_2 z_e'' \right) dt \tag{7.39}$$

A Ljapunov function shall now be constructed: Let $z'' = z_r'' + Q_1 z_e' + Q_0 z_e$ be a solution of a differential equation with

$$Q_1 = \begin{pmatrix} q_{1x} & 0 \\ 0 & q_{1y} \end{pmatrix} \qquad Q_0 = \begin{pmatrix} q_{0x} & 0 \\ 0 & q_{0y} \end{pmatrix} \tag{7.40}$$

$q_{1x}, q_{1y} > 0$ and $q_{0x}, q_{0y} > 0$; R_1 and R_2 are positive definite and symmetric. $\Longrightarrow (z_e, z_e')$ is a stable equilibrium state.

A proof runs over the construction of a Ljapunov function as a quadratic form: Let $V(z_e, z_e') = 1/2(z_e^T Q_0 z_e + z_e'^T z_e)$ then show it fulfills the requirements of a Ljapunov function

- $[z_e, z_e'] = 0 \longrightarrow V(z_e, z_e') = 0$

- $V(z_e, z_e') = 1/2((q_{0x}x_e^2 + q_{0y}y_e^2) + (x_e'^2 + y_e'^2))$
  $V(z_e, z_e') > 0$ for all $[z_e, z_e'] \neq 0$ because of $q_{0x}, q_{0y} > 0$

- $V'(z_e, z_e') = z_e'^T Q_0 z_e + z_e'^T z_e''$
  because of $z_e'' + Q_1 z_e' + Q_0 z_e = 0$
  $V'(z_e, z_e') = z_e'^T Q_0 z_e + z_e'^T (-Q_1 z_e'' - Q_0 z_e)$
  $V'(z_e, z_e') = -z_e'^T Q_1 z_e'.$

Now from construction q_{1x} and $q_{1y} > 0$
$\longrightarrow V'(z_e, z_e') = -z_e'^T Q_1 z_e' \leq 0$ for all $[z_e, z_e']$
$\longrightarrow V'(z_e, z_e') = 1/2(z_e^T Q_0 z_e + z_e'^T Q_1 z_e')$ is a Ljapunov function.
$\longrightarrow (z_e, z_e')$ is globally asymptotically stable as depicted in figure 7.58.

The vehicle following the dashed curve finally catches up the reference point running along the black reference path.

Figure 7.58 Convergence of the vehicle movement according to the planned trajectory defined by the reference point

Calculating $x(t)$ and $y(t)$ As x_r'', y_r'' and z_e and z_e' are given, the differential equations for $x(t)$ and $y(t)$ can be solved

$$x'' = x_r'' + q_{1x}(x_r' - x') + q_{0x}(x_r - x) \qquad\qquad q_{0x} > 0 \, q_{1x} > 0 \quad (7.41)$$
$$y'' = y_r'' + q_{1y}(y_r' - y') + q_{0y}(y_r - x) \qquad\qquad q_{0y} > 0 \, q_{1y} > 0 \quad (7.42)$$

These are linear second order differential equations. Their solutions are damped oscillations. Of interest are solutions near the aperiodic limit.

Let $v_{\max} =$ be the maximum value of the control function and $v_{\max}' =$ the maximum value of the velocity of the control function then choose $q_{0x} = q_{0y} = 0{,}01(v_{\max}'/v_{\max})^2$ and $q_{1x} = q_{1y} = 2 \cdot \xi \cdot 0{,}1(v_{\max}'/v_{\max})$ with $\xi = 0{,}7 \ldots 1{,}0$. Figure 7.59 shows the development of $x(t)$ for the asymptotic limit $\xi = 0{,}7$ and a value slight out of it. In both cases $x(t)$ runs towards x_r.

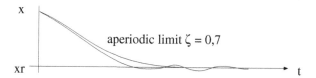

Figure 7.59 Convergence of $x(t)$ towards x_r for two values of ξ

Global stable path by Ljapunov function The criteria for a Ljapunov function were

- $V(z,z') = 0$ for $(z,z') = 0$

- $V(z,z') > 0$ for $(z,z') \neq 0$

- $V'(z,z') \leq 0$ for all (z,z')

$\Longrightarrow (z,z')$ asymptotically stable:

$$\lim_{t\to\infty} z(t) = 0 \lim_{t\to\infty} z'(t) \quad = 0V(z,z') = V(x(t),y(t),x'(t),y'(t)) = V(x,y;t)$$
$$(7.43)$$

Figure 7.60 shows the situation: the third criterion is sketched on the left and in the middle of the figure: going from $t \longrightarrow t + \Delta t$, $V(x,y;t)$ shrinks. On the right the first and second criterion are sketched: at $(z,z') = 0$ the value of V is zero. For all $(z,z') \neq 0$ $V > 0$, but shrinking towards $V = 0$ over t.

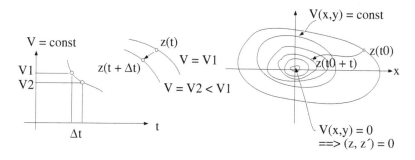

Figure 7.60 Development of $V(z,z')$

8 Control Architectures

The generation of designated robot behavior is one of the most difficult problems when designing the control system for robotic applications with many sensors and actuators. Due to the diversity of tasks an autonomous vehicle has to fulfill, the control has to be embedded into a convenient framework. The process of building up a control system should be supported by an adequate methodology to help overcoming difficulties common to complex robotic systems, e.g. ensuring secure operation, modularity, or handling a system of growing complexity. Therefore, different types of control architectures have appeared with contrary approaches for tackling the emerging problems.

A control architecture is a framework enabling a system to fulfill the following tasks:

Fusion of sensor data into logical sensors Sensor data has to be preprocessed for usage for localization, generation of obstacle maps, generation of maps for navigation and planning, object recognition, display and representation of knowledge.

Motor controller Access to the hardware has to be provided by the realization of a convenient motion control interface, e.g. velocity v and angular velocity ω.

Pilot function A Pilot realizes the path control via specification of motion commands which are required for e.g. collision avoidance, driving through narrow passages, turning in dead-end situations.

Navigation The navigator calculates accessible tracks to be traversed via the pilot function. The plan includes avoidance of obstacles and requires knowledge about the surrounding area.

Planning function A planning component generates actions and sets targets for the navigation component. This includes strategic decisions and keeping the overview concerning given tasks.

User interaction Access to the control software has to be provided by a suitable Human-Machine-Interface (HMI).

For all tasks mentioned above world knowledge has to be considered, which has to be provided adequately.

In general, control architectures in the field of robotics can be distinguished into hierarchical vs. distributed and task-oriented vs. behavior-based (see figure 8.1).

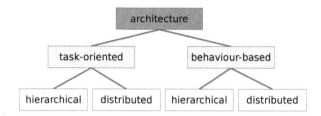

Figure 8.1 Overview over control architectures commonly used in the field of robotics

Hierarchical architectures depend on the assumption that tasks can be divided into subtasks which are arranged so that higher level components generate subgoals for lower level components. In contrast to this, distributed architectures allow the assignment of subtasks to independent components. A suitable communication mechanism is required for data transfer between the parts involved.

The distinction between task-oriented and behavior-based control architectures is reflected in the decomposition of a given task. Task-oriented architectures depend on a central world model which is manipulated and evaluated by the different components (sensing, modeling, planning, execution). The tasks of processing sensor data as well as generating control values for the robot are encapsulated into responsible components with exclusive access. Behavior-based control architectures, however, are designed by decomposing a given task into independent behaviors, each of which keeps its own compact representation of the environment which is required for task execution. Here all components have unlimited access to sensor data and to the control interface. This, however, requires mechanisms for the coordination of conflicting data.

Besides the given distinction of control architectures into task-oriented vs. behavior-based and hierarchical vs. distributed, [Mat97] emphasizes the difference of architectures in respect to the degree of deliberation, see figure 8.2. Reactive architectures not supporting knowledge storage are contrasted with deliberative architectures with an elaborate model of the world. The former supports immediate reaction on occurring situations but has weaknesses in respect to planning tasks requiring a certain degree of memory. The latter is suited to complex tasks but has drawbacks concerning outdated knowledge, false sensor readings, or reaction time.

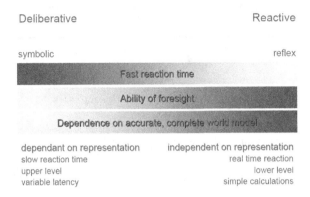

Figure 8.2 Properties of deliberative and reactive control

In order to combine the advantages of both characteristics, hybrid architectures have emerged with a lower reactive layer and a higher deliberative one. This, however, breaks the control system into two inhomogeneous parts with different characteristics. In contrast to this, behavior-based control systems store a representation of the environment which is distributed among the single components and therefore can combine reactive and deliberative components into one architectural design.

In the following, a task-oriented as well as several behavior-based control architectures are discussed in more detail.

8.1   The hierarchical task-oriented control architecture RCS

As an example of a hierarchical task-oriented control architecture, the Real-Time Control System (RCS) by James S. Albus [Alb92] is presented here. Since its beginning in the 1970s at the NIST (National Institute of Standards and Technology), the architecture has constantly advanced and been applied to several projects, e. g. the DEMO III project [LMD02].

RCS is composed of hierarchically arranged layers consisting of one or more RCS nodes. As depicted in figure 8.3, each of them has 4 modules: Behavior Generation (BG), World Modeling (WM), Sensory Perception (SP) and Value Judgment (VJ). Additionally, the Knowledge Database (KD) and the Operator Interface (OI) are provided.

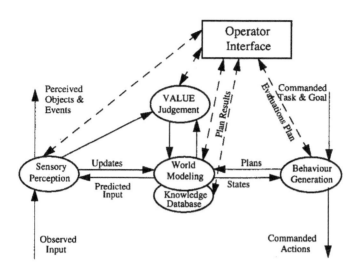

Figure 8.3 A RCS-4 node and the data flow between the included modules [Alb92]

The BG module plans and controls the actions of the system. For this purpose a complex plan is decomposed into simpler tasks by using the information provided by the higher layer as well as WM and VJ of the same layer.

The SP module processes sensor data and compares them with predictions of the WM module. Additionally, SP perceives objects, events, and situations and transfers them to the WM module.

The WM module uses information given by SP to update the Knowledge Database. Predictions of sensor values as well as the simulation of plans proposed by the BG module are additional tasks of this module.

The VJ module calculates cost, risk, and benefit of simulated plans. It distinguishes between important and irrelevant objects and events and transfers the results to the BG module.

The Knowledge Database contains data about the environment. Each of the modules has access to the KD, either directly or indirectly via the WM module.

Finally, the OI is the interface for observing or manipulating the system behavior by a human operator.

RCS supports two kinds of communication. On the one hand, communication between RCS nodes of different layers involves the transfer of new tasks to the BG module of the lower layer and newly perceived objects and events to the SP module of the higher layer. On the other hand, modules inside a RCS node communicate as depicted in figure 8.3.

This results in two main data flow directions: One top down information flow consisting of tasks and goals between the BG modules and one bottom up information flow consisting of processed sensor data. Due to the growing complexity of higher layers, the higher a layer is, the longer the cycle takes. An example for a timing diagram is depicted in figure 8.4

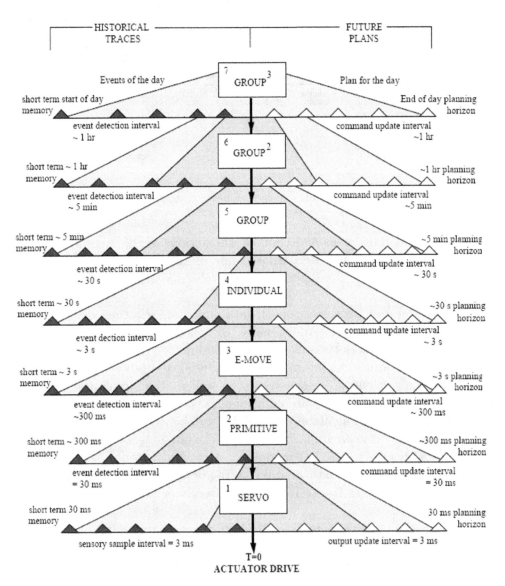

Figure 8.4 Example for the time horizon of different layers in RCS [Alb92]

Here, seven layers are covering a time range between milliseconds and days. The horizontal axis consists of historical traces (left) and future plans (right). Both of them are processed within the same time scale.

While the applicability of hierarchical task-oriented control architectures like RCS was shown in several applications, some disadvantages have emerged. At first the control depends on a consistent world model. However, this central representation of the environment is prone to errors due to false sensor readings or outdated information. Here the immediate usage of sensor data leads to a more responsive and correct behavior.

Second, the functionality of the whole system depends on the proper operation of all components. A different approach, in which malfunctions of single modules can be caught by others is the behavior-based approach, presented in the next section.

Finally, hierarchical task-oriented control architectures tend to limit the extensibility of the system as a new functionality is reflected in the change of a multitude of existing components. Therefore, the scalability of these systems is limited.

8.2   Behavior-based control architectures

The development of robotics was heavily influenced by paradigms in Artificial Intelligence (AI) like the following (Marvin Minsky [MMN55]): "[An intelligent machine] would tend to build up within itself an abstract model of the environment in which it is placed. If it were given a problem it could first explore solutions within the internal abstract model of the environment and then attempt external experiment." Therefore, knowledge representation, planning reasoning and hierarchical composition were the main fields of research reflecting the human understanding of intelligence (e. g. STRIPS [FN71], ABSTRIPS [Sac74], NOAH [Sac75]).

Inspired by observations in biology, these traditions were questioned. According to Brooks, "planning is just a way of avoiding figuring out what to do next" [Bro87]. Therefore, a shift in paradigm took place, moving from sensing and acting to behavior-based robotics where preferably simple agents show intelligence through coordinated behavior.

The motivations for this change were the following:

- Complex behavior does not necessarily arise from complex control systems.

- The real world is the best model.

- Programming should be kept simple.

- Systems should show robustness at noisy sensor readings.

- Systems should provide the possibility of incremental design.

- All calculations should be performed on-board and therefore fit the machine time available.

The motivation for the change from task-oriented to behavior-based control would be best expressed by Thomas Huxley: "The great end of life is not knowledge, but action".

The following examples of behaviors could appear in wheel-driven or legged mobile autonomous robots:

- Exploration (movement in a general direction)

- Targeted (movement in direction of attractors)

- Avoidance (avoid collision with objects and environment)

- Path following (e. g. wall, planned path, stripe)

- Posture control (balance, stability)

- Social behavior (e. g. parting, hives, flocks)

- Remote-autonomous behavior (user interaction, coordination)

- Perceptual behavior (visual search, ...)

- Walking behavior

- Manipulator behavior, grip behavior

In contrast to task-oriented control architectures, behavior-based approaches have proven to handle emerging difficulties rather well. They do not depend on the correctness of a central world model, make it easy to incrementally add functionality while handling increasing complexity and show robustness to unknown sensor data due to an overall functionality emerging from the interaction of multiple generalizing behaviors.

Still, the problem of controlling complex robotic systems is not solved by the behavior-based paradigm alone. Rather, while helping with some common problems, behavior-based architectures introduce new difficulties. Among those is the question of how to coordinate multiple and possibly competing behaviors running in parallel and trying to act on the same actuators. Another issue is the identification of error sources in a control that shows an emergent system behavior rather than an explicitly implemented one. Also, there is the matter of how the architecture can help structuring the design process, e.g. giving support in the process of selecting the best set of behaviors and coordinating their action.

Several variants of behavior-based architectures have been developed in order to tackle the presented problems. In the following section, the Subsumption Architecture by R. Brooks [Bro86], reuse and temporal sequences of behaviors by Nicolescu and Mataric [NM00, NM02], and the iB2C[1] architecture of the Robotics Research Lab in Kaiserslautern are presented. Other behavior-based architectures include R. Arkin's works [Ark89, Ark98] on schema-based and potential field approaches, further methods by M. Mataric [Mat92, Mat97], miscellaneous fuzzy approaches [LRM94, SKR95, KS97, Ros97, SDC05], the Dynamical System Approach by Althaus/Christensen [AC02], Behavior Oriented Design [Bry01], parallel behavior execution without action selection mechanism [Ste94], or activation-based [BILM03, CA07] as well as neural-network-based architectures [FM96, Bee96].

8.2.1 The Subsumption Architecture

The first architecture implementing the presented ideas was the Subsumption Architecture developed by Rodney Brooks, MIT, 1986 [Bro86]. He proposed the alignment of behaviors along horizontal layers, with all behaviors having access to all sensor data to generate actions for all actuators. The interaction of behaviors is carried out through interdiction of inputs and overruling of outputs. The primary feedback between behaviors occurs via the environment. Using the special C derivative **interactive C** allows for system state analysis during runtime. The approach was applied to several mobile vehicles as well as to the walking machines Genghis and Attila.

This approach has many advantages. Inherently, the architecture supports running behaviors asynchronously in parallel while data connections can be prone to errors and delays. Due to the independence of the components, extending the system with sensors and behaviors is directly supported.

---

[1] iB2C: integrated behavior-based control

The implementation of the individual modules is kept simple and the system shows a certain robustness in case of malfunction of behavior modules or behavior levels. Finally, simultaneously achieving multiple goals can be easily implemented.

The structure of the basic elements of the Subsumption Architecture is shown in figure 8.5. Each of them has input and output lines. Via a suppressing mechanism, input signals can be replaced by the suppressing signal given into the circle. Output signals of modules can be inhibited, i.e. any output signal is blocked for the given time. Each module internally implements a finite state machine which can be reset to state NIL via the corresponding input.

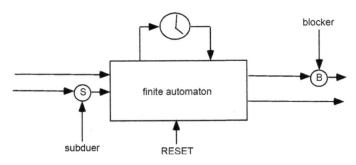

Figure 8.5 Basic module of the Subsumption Architecture

The system design follows the structure given in figure 8.6. Its control is split up into layers with higher layers subsuming the functionality of lower level layers when desired. The system can be partitioned at any layer so that the lower layers form a complete operational system. An example of the functionality of layers is depicted in figure 8.7.

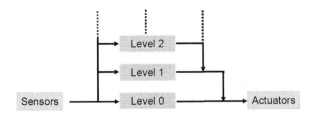

Figure 8.6 Layout of the behavior levels in the Subsumption Architecture

This architecture has proven suitable for small systems and has successfully managed navigation in an office environment with many obstacles. However, this approach tends to run into scalability problems due to limitations concerning the amount of internal representation. Also, the reuse of components is often not possible and weaknesses occur when behaviors are added to an existing system. Therefore, further enhancements have been undertaken in the years after the introduction of Subsumption Architecture.

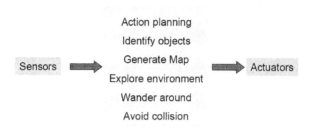

Figure 8.7 Example for the functionality of layers in the Subsumption Architecture

8.2.2 Reuse and temporal sequences of behaviors

The work of Nicolescu and Mataric (see [NM00, NM02]) identifies common deficiencies of behavior-based architectures: the difficulties in reusing existing behaviors for new tasks, the inability to easily realize temporal sequences of behavior activations, and the lack of support for the automatic creation of behavior networks.

The first problem is addressed by splitting up a behavior into an "abstract" and a "primitive" part, with the former constituting the interface of a behavior and the latter containing the actual functionality (see figure 8.8). The abstract part also contains checks to ensure that the preconditions of the behavior are fulfilled. So when using a behavior for different tasks with different preconditions, the primitive part can be reused and only the interface part has to be altered.

Behavior networks can be created by connecting abstract behaviors. This can either be done by hand or by an algorithm which analyzes the preconditions and effects of behaviors and uses backtracking to create a network for a specific task. Dealing with large networks containing many behaviors is facilitated by the support for grouping behaviors and thus building hierarchical networks. The architecture provides different types of connections between behaviors, allowing the creation of temporal sequences. By encoding task-specific aspects into behavior connections instead of directly into the behaviors, behaviors can be built in a more general way and thus be reused more easily.

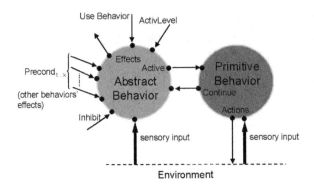

Figure 8.8 An abstract and a primitive behavior and the connection between the two [NM02].

[NM03] describes an extension of the architecture that allows robots to learn tasks from a robot or human teacher. During a demonstration phase, the learning robot makes observations and logs which of his behaviors can be used to achieve a certain situation. In a second phase, it uses these observations to build a behavior network for fulfilling the demonstrated task. So what is learned is when to use which behavior. New behaviors are not learned.

8.3   The integrated behavior-based control architecture iB2C

As an example for a behavior-based control architecture which addresses the issues mentioned before, the iB2C architecture of the Robotics Research Lab (University of Kaiserslautern) is presented here [PLB10]. The goal is to find a behavior-based architecture of modular structure with control units ranging from motor schemes up to deliberative planning layers. Common interfaces help in reusing and easily adding modules, and the arbitration mechanism allows for separating the control data from the coordination data flow.[2] The architecture is applicable to a wide range of robotic systems. It also shows design guidelines to simplify the creation of a consistent, robust and maintainable system. Last but not least, a programming framework supports the implementation by providing suitable tools for designing, debugging and

---

[2] Here control data is referred to as values for actuators, e. g. velocity, while the coordination data flow provides information about behavior states and is used for the internal behavior interaction of the system.

inspecting a control system of growing complexity. The proposed architecture is a further development of the behavior-based control as previously introduced in [ALBD03], where it has mainly been used to control walking machines.

8.3.1   The basic behavior module

The fundamental unit of the proposed control architecture is the behavior module (see figure 8.9) which is based on [Alb07] and [PLB05]. Each atomic behavior is wrapped into such a module with a uniform interface.

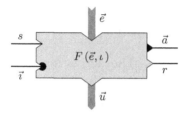

Figure 8.9   Basic iB2C behavior module

Behaviors can be described as three-tuples of the form

$$B = (f_a, f_r, F) \tag{8.1}$$

where f_a is the *activity function*, f_r is the *target rating function*, and F is the *transfer function* of the behavior. These functions generate *activity* information $\vec{a}$, a *target rating* r, and an *output vector* $\vec{u}$, respectively. Additionally, each behavior receives an *input vector* $\vec{e}$, a *stimulation* s, and an *inhibition vector* $\vec{\imath}$. In the following, these characteristics are explained in more detail.

Behaviors receive data required for fulfilling their work via the *input vector* $\vec{e} \in \mathbb{R}^m$ which can be composed of sensory data (e. g. distance measurements) or information from other behaviors (e. g. their target rating). The *output vector* $\vec{u} \in \mathbb{R}^n$ transmits data generated by the behavior (e. g. intended velocity values). This output describes the data which is used for actuator control or as input for other behaviors.

Each behavior provides standardized inputs for adjusting its relevance:

Definition 8.1 (Stimulation). The stimulation $s \in [0,1]$ of a behavior B is an input determining the intended relevance of B. In this notation, $s = 0$ indicates no stimulation and $s = 1$ a fully stimulated behavior. Values between 0 and 1 refer to a partially stimulated behavior.

Stimulation can be used to adjust the influence of competing behaviors or to allow higher-level behaviors to recruit lower-level behaviors and their functionality by explicitly stimulating them. Certain behaviors require constant stimulation, e. g. safety behaviors or reflexes. These behaviors are depicted by a filled triangle at the stimulation port in the figures.

Definition 8.2 (Inhibition). Each behavior can be inhibited by k other behaviors via its input $\vec{i} \in [0,1]^k$. The inhibition $i \in [0,1], i = \max\limits_{j=0,...,k-1}(i_j)$ of a behavior B reduces the relevance of B. Here $i = 1$ refers to full inhibition, $i = 0$ to no inhibition. Values between 0 and 1 refer to a partially inhibited behavior.

Therefore, inhibition has the inverse effect of stimulation.

Definition 8.3 (Activation). The activation ι of a behavior B indicates the effective relevance of B in the behavior network. It is composed of the stimulation s and the inhibition i, with

$$\iota = s \cdot (1 - i) \tag{8.2}$$

The calculation of the outputs of a behavior is implemented by the transfer function F, the activity function f_a, and the target rating function f_r. The transfer function $F(\vec{e}, \iota)$ determines the output vector $\vec{u}$, where

$$F : \mathbb{R}^m \times [0,1] \rightarrow \mathbb{R}^n, \quad F(\vec{e}, \iota) = \vec{u} \tag{8.3}$$

F provides the intelligence of a behavior, calculating actions depending on input values and internal representations. This can be a reactive response to input values, but also a more complex calculation like a state machine or sophisticated algorithms. This way, both reflexive sensor-actor coupling and deliberative behaviors can be implemented (as postulated for behavior-based architectures by [Mat97]).

Each behavior provides two behavior signals that allow for deducing information about its state and its assessment of the current situation:

Definition 8.4 (Activity). The behavior signal activity $a \in [0,1]$ of a behavior B represents the amount of influence of B in the current system state. $a = 1$ refers to a state where all output values are intended to have highest impact, whereas $a = 0$ indicates an inactive behavior. Values between 0 and 1 refer to a partially active behavior.

The activity a and the derived activities $\vec{\underline{a}}$ are defined by the activity function f_a with

$$f_a : \mathbb{R}^m \times [0,1] \to [0,1] \times [0,1]^q, \; f_a \left(\vec{e},\iota\right) = \vec{a} = (a, \underline{\vec{a}})^T \tag{8.4}$$

where

$$\underline{\vec{a}} = \left(\underline{a}_0, \underline{a}_1, \ldots, \underline{a}_{q-1}\right)^T \tag{8.5}$$

with

$$\underline{a}_i \leq a \quad \forall i \in \{0,1,\ldots,q-1\} \tag{8.6}$$

The derived activities $\underline{\vec{a}}$ allow a behavior to transfer only a part of its activity to other behaviors.

Definition 8.5 (Target rating). The behavior signal target rating $r \in [0,1]$ is an indicator for the contentment of a behavior. A value of $r = 0$ indicates that the behavior is content with the actual state, while $r = 1$ shows maximal dissatisfaction. Values between 0 and 1 refer to a partially content behavior.

To ensure a consistent behavior network during the development process, some principles have to be complied with. Similar to [HA01] these principles allow some basic assumptions about the structure of the control system. These are required for the analysis of system properties.

As the activation defines the upper bound of a behavior's influence, the following principle must be observed:

Principle 8.1 (Activity limitation). The activity a of a behavior B is limited by the activation ι of B: $a \leq \iota$

Furthermore, if the system is in the goal state of a behavior (characterized by $r = 0$), it intends to maintain its adjusted influence. Therefore, the following principle is postulated:

Principle 8.2 (Goal state activity). The activity a of a behavior B does not change in case $r = 0$ and $\iota = \text{const}$.

Usually a behavior's activity is $a = 0$ in case it is situated in its goal state, but there are cases where a constant influence is required, i.e. $a > 0$. An example is a behavior generating torque for an arm joint. If, in this case, the behavior's activity was lowered in the goal state, external forces or competing behaviors could change the adjusted joint angle.

In contrast to the influence of the activation on the activity, the target rating only depends on the input vector and the behavior-internal state. This way, the target rating is an indicator for a behavior's state assessment, leaving out external adjustments of its influence:

Principle 8.3 (Target rating independence). There is no (direct, i. e. inside a behavior) influence of the activation ι on r.

As described before, behavior-based architectures do not work with a centralized world model. This is represented by the fact that actions of a behavior only depend on the input vector $\vec{e}$, their activation and the behavior-internal representation of the current situation, which can be non-existent for certain behaviors.

Example behavior **Turn to object**

In order to exemplify the calculation of the described behavior properties, this section describes a showcase behavior rotating a vehicle to a detected object in front. As input vector $\vec{e}$ the behavior receives the angle β to the object to be followed. The output $\vec{u}$ is a normalized rotation value rot $\in [-1,1]$. As the rotation output shall point the robot into the direction of the object, the transfer function can be defined as:

$$\text{rot} = \begin{cases} -1 & \text{if} \quad \beta < -\beta_{\max} \\ \frac{\beta}{\beta_{\max}} & \text{if} \quad -\beta_{\max} \leq \beta \leq \beta_{\max} \\ 1 & \text{if} \quad \beta > \beta_{\max} \end{cases} \qquad (8.7)$$

The target rating indicates the contentment of the behavior with the current situation. As the goal is to point the vehicle into the direction of the object, the behavior becomes discontent according to the angle to the object:

$$r = h(\beta) \qquad (8.8)$$

with

$$h(\beta) = \begin{cases} \frac{|\beta|}{\beta_{\max}} & \text{if} \quad |\beta| \leq \beta_{\max} \\ 1 & \text{else} \end{cases} \qquad (8.9)$$

As the behavior intends to reduce the deviation to the object, its activity has to increase if the angle to the object grows. The activation ι limits a in order to meet Principle 8.1:

$$a = \iota \cdot h(\beta) \qquad (8.10)$$

8.3.2   Fusion behavior module

A behavior-based system certainly is not completed with the implementation of the single behaviors. As the influence of behaviors on control values or on other behaviors interleaves, and as they can have contrary goals, their outputs must be usefully combined. This question of behavior coordination is often considered the main problem in developing such an architecture.

The behavior coordination within iB2C networks is achieved by so-called *fusion behaviors* (see figure 8.10). These are integrated in the case of competing behaviors.

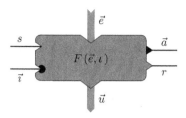

Figure 8.10 Fusion behavior module in iB2C

Fusion behaviors have the same interface as defined by the basic behavior module. For the coordination of p competing behaviors B_c, the input vector $\vec{e}$ is composed of

- the activities a_c (or the derived activities $\underline{a}_c^i$ of the vector $\underline{\vec{a}}_c$ respectively),

- the target ratings r_c, and

- the output vectors $\vec{u}_c$.

The transfer function F is the fusion function processing input values to a merged output control vector $\vec{u}$.

An example of the fusion of three competing behaviors B_c, $c \in \{0,1,2\}$ is depicted in figure 8.11. Each of the B_c is connected to the fusion behavior by its behavior signals a_c and r_c as well as the output vector $\vec{u}_c$. For clarification, the input vector of the fusion behavior is drawn separately.

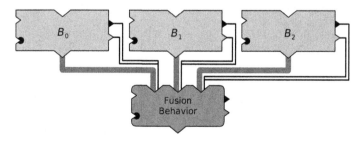

Figure 8.11 Exemplary fusion of three behavior outputs

The underlying assumption of the fusion of output values is that behaviors with a high activity deserve a higher influence on the control output than those with a lower activity. By using the behavior signal *activity* as a means for coordinating the behaviors, the control data flow and the coordination data flow are separated.

The behavior signal calculation of fusion behaviors has to comply with the following principle:

Principle 8.4 (Fusion behavior neutrality). The calculation of the activity a and the target rating r of a fusion behavior must keep the following conditions:

$$\min_c(a_c) \cdot \iota \le a \le \min\left(1, \sum_{j=0}^{p-1} a_j\right) \cdot \iota \tag{8.11}$$

$$\min_c(r_c) \le r \le \max_c(r_c) \tag{8.12}$$

This way, it is guaranteed that a fusion behavior does not inject or remove activity, as expected from a coordination component. Furthermore, there is no improvement or deterioration of satisfaction. This accounts for the fact that calculations concerning the assessment of state are only located in non-fusion behavior modules.

The following sections describe the set of fusion function implementations being used.

Maximum fusion (winner takes all)

In the case of maximum fusion the control value of the most active behavior is forwarded. Other behaviors obtain no influence. The transfer function F is defined as:

$$\vec{u} = \vec{u}_s \text{ where } s = \operatorname*{argmax}_c(a_c) \tag{8.13}$$

Activity and target rating are set according to the most active behavior:

$$a = \max_c(a_c) \qquad r = r_s \text{ where } s = \operatorname*{argmax}_c(a_c) \tag{8.14}$$

The maximum fusion implements a switching between behaviors and is suitable when a combination of control outputs leads to unwanted results.

Weighted fusion

In the case of weighted fusion the control values of the competing behaviors are weighted with the activity of the corresponding behavior. This way, a subtle gradation of coordinating behavior control outputs regarding their activity is achieved.

The transfer function F is defined as:

$$\vec{u} = \frac{\sum\limits_{j=0}^{p-1} a_j \cdot \vec{u}_j}{\sum\limits_{k=0}^{p-1} a_k} \tag{8.15}$$

The activity is defined as:

$$a = \frac{\sum\limits_{j=0}^{p-1} a_j^2}{\sum\limits_{k=0}^{p-1} a_k} \cdot \iota \tag{8.16}$$

The target rating of a fusion behavior indicates its goal to satisfy highly activated input behaviors and is calculated as follows:

$$r = \frac{\sum\limits_{j=0}^{p-1} a_j \cdot r_j}{\sum\limits_{k=0}^{p-1} a_k} \tag{8.17}$$

Weighted sum fusion

The weighted sum fusion is used for summing up the control values of the competing behaviors according to their activity. Applications for this fusion function are cases where several behaviors contribute to a torque of a joint or in cases where vectors are added up.

The transfer function F is defined as:

$$\vec{u} = \sum_{j=0}^{p-1} \frac{a_j \cdot \vec{u}_j}{\max\limits_c (a_c)} \tag{8.18}$$

The activity is defined as:

$$a = \min \left(1, \sum_{j=0}^{p-1} \frac{a_j^2}{\max\limits_c (a_c)} \right) \cdot \iota \tag{8.19}$$

The target rating calculation is the same as for the weighted fusion:

$$r = \frac{\sum\limits_{j=0}^{p-1} a_j \cdot r_j}{\sum\limits_{k=0}^{p-1} a_k} \tag{8.20}$$

8.3.3 Behavior interaction

Besides communication between behaviors through the environment or by using arbitrary data, the behavior interaction mainly takes place by transferring activity data between behaviors. As activity defines the relevance of behaviors and their outputs, the transfer inside the behavior network is restricted as follows:

Principle 8.5 (Stimulation/inhibition restriction). Inside iB2C behavior networks a behavior B may only be stimulated or inhibited by the activity a or $\underline{a}_i$ of the vector $\vec{a}$ of other behaviors.

This way, it is clearly defined where activity is injected into the behavior network and how it is transferred to other behaviors. Consequently, it is impossible for a behavior to gain influence without being sufficiently stimulated. The flow of activity through iB2C networks therefore allows statements about the overall system behavior.

In contrast to the activity signal, the target rating of behaviors counts as local situation assessment. It is used as abstract sensor value or for evaluating regions of dissatisfaction but is not transferred through the whole behavior network.

Due to the importance of the behavior signals activity and target rating, these values must be present everywhere inside the behavior network:

Principle 8.6 (Behavior signal availability). Each control value entering a behavior network must be provided with a suitable activity and target rating value. Activity and target rating must not be dropped until control values leave the behavior network, i. e. until they are transformed to actuator commands.

This principle guarantees that for each control value an assessment of the relevance is provided. This is a key aspect which allows further processing of the data in the behavior network.

More precisely, the following sources and sinks of activity can be specified:

Sources of activity: Activity can enter the behavior network as follows:

1. Behaviors are stimulated from outside the behavior network.

2. Behaviors are constantly stimulated.

3. The activity of a behavior is used as stimulation for several other behaviors.

Sinks of activity: Activity can be reduced inside the behavior network as follows:

1. Fusion Behaviors combine several input activities to one output activity.

2. An activated behavior can emit an activity $a < \iota$ or a derived activity $\underline{a}_i < \iota$ of the vector $\underline{\vec{a}}$.

3. A behavior which is inhibited reduces the amount of activity at that place in the network.

4. If a behavior's activity output is not connected to another behavior, its activity is lost (e. g. in case a control value leaves the behavior network).

The previously defined principles allow the usage of the flow of activity for deriving each behavior's influence on other parts of the behavior network.

8.3.4 Behavior coordination

The behavior-based approach implies that several behaviors can contribute to the same control value. Therefore, the coordination of behaviors requires suitable mechanisms. This is where behavior architectures differ most. The following shows how a multitude of coordination mechanisms can be implemented in iB2C based on the uniform behavior module model (including the fusion behaviors).

A distinction of mechanisms for behavior coordination is presented in [Pir99]. Here, the first criterion distinguishes if several behaviors are *arbitrated*, i. e. one behavior or a set of them has control for a period of time, or if their *commands are fused*, i. e. a combination of control outputs of the behaviors takes place.

Arbitration makes sense when behavior actions have to be transferred without modification. The following types can be distinguished [Pir99]:

Priority-based mechanisms: Behaviors are selected according to priorities assigned to each of them (e. g. [Bro86]).

Priority-based arbitration in iB2C is implemented using inhibition of behaviors, see figure 8.12. The order of the behaviors determines the priority of each component. The maximum fusion behavior selects the most active behavior.

State-based mechanisms: Behaviors are selected in respect to the current state and the competence of behaviors for handling the situation (e. g. [KCB97]).

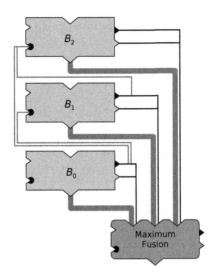

Figure 8.12 Priority-based arbitration in iB2C

State-based arbitration is realized using a behavior containing state evaluation mechanisms which stimulates action generating behaviors. Coordination takes place using a maximum fusion behavior.

If the state evaluation relies on feedback of the action generating behaviors, the activity and the target rating of the respective behaviors can be used.

Winner-takes-all mechanisms: One of the behaviors is selected as a result of a competition between them (e. g. [Mae89]).

The Winner-takes-all mechanism is directly supported in iB2C by the maximum fusion. Here, the competition between the behaviors is implemented as activity calculation.

In contrast to arbitration, command fusion supports the combination of behavior outputs. Several solutions for representing the desired commands and for determining the relevance of commands have been developed [Pir99]. Besides command fusion using the weighted sum fusion function, iB2C directly supports the superposition and voting mechanisms.

Voting: Each behavior votes for different actions. After combining them, the action receiving the highest number of votes is chosen (e. g. [PRK91]).

In iB2C, voting is implemented using a standard fusion behavior and a mapping behavior (see figure 8.13). Each behavior involved provides votes for each of the n possible options (e. g. driving directions) which are transferred to the fusion behavior implementing the weighted fusion function. The output of the fusion behavior consists of the weighted votes for each voting option. A mapping behavior stimulated by the fusion behavior then maps the maximal option rating to a command for further processing.

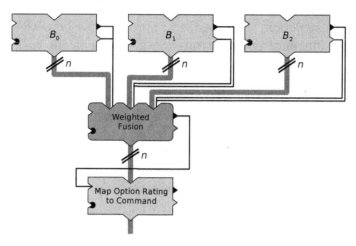

Figure 8.13 Voting mechanism in iB2C

Superposition: Behavior actions are represented as vectors which are linearly combined (e. g. [Ark87]).

Superposition in iB2C is implemented by the weighted sum fusion, where a component-wise fusion takes place with the activity representing the relative scale of each vector.

Fuzzy: Similar to voting mechanisms, here fuzzy inferencing techniques are used (e. g. [SKR95]).

Multiple objective: Also similar to voting, the desirability of actions is defined by each behavior's objective function. Coordination is carried out by looking for actions that sufficiently satisfy all objective functions by using multiple objective decision theory methods (e. g. [PHTO$^+$00]).

As fuzzy inferencing techniques and multiple objective mechanisms implement functionality similar to voting, they are not treated here.

8.3.5   Design guidelines

Designing a control system for robotic applications requires a systematic methodology in order to cope with the complexity of sensor processing and

control data generation. In iB2C, the development begins by figuring out the
relevant degrees of freedom (DOF), e. g. rotation and velocity of a vehicle,
emotional actuators of a humanoid head, or joint motions of legs. Each of the
DOF is divided into positive and negative direction, leading to two control
data paths for every motion possibility. The conflation of the data flow is
accomplished using a fusion behavior for each of the DOF. Depending on the
mechanical construction, the described approach may be performed several
times for each kinematic chain involved, e. g. for a pan tilt unit of a camera
head or for a multitude of legs.

In order to fulfill basic safety requirements, the next step is introducing
behaviors reacting on safety related sensor input (e. g. stopping or turning
away a vehicle because of data provided by a proximity sensor). Each of the
safety behaviors influences a DOF by using its activity output for inhibiting
fusion behaviors of the layer above and by propagating a new command to a
fusion behavior in the layer below. As each of the DOF is divided into positive
and negative direction, behaviors can be integrated in such a way that only
the supervised direction is influenced.

This procedure results in an interface for higher level behaviors and
encapsulates the functionality of a safety behavior system. High-level behaviors are then added using a top-down task-oriented approach. Methods
like those proposed in [Bry01], asking for *what* to do *how* and *when* and
iteratively revising the structure can be applied here.

Hierarchical abstraction   One advantage of the decomposition of tasks
into behaviors is the low complexity of each behavior. However, the result
of this approach often is a network with a large number of behaviors. In
order to simplify the structure and to clarify the functionality, a hierarchical
abstraction becomes necessary. In the case of iB2C this can be accomplished
using behavioral groups (see figure 8.14). These are groups in the sense of the
embedding programming framework[3], i. e. a collection of modules or further
groups with a new interface and dedicated connections between group and
modules. A behavioral group acts as a new behavior, providing the same
standardized input and output signals described in section 8.3.1.

The challenge for the developer is finding sets of behaviors representing
new semantic groups. One approach is to reflect the implemented decomposition in the hierarchical structure of groups. Another hint for grouping
behaviors stems from the influence of multiple behaviors on a DOF. If several
behaviors work in the same domain and have an influence on the same data

---

[3] in this case the *modular control architecture* MCA, [SAG01]

path in the network (e. g. behaviors using different sensor systems for bringing a vehicle to a halt), these behaviors are good candidates for forming a new group.

When constructing a behavior network, the designer has to question himself about the semantics behind behaviors and whether a group of behaviors separates from the individuals to form a new semantic unit. If this is the case, a new behavioral group should be introduced.

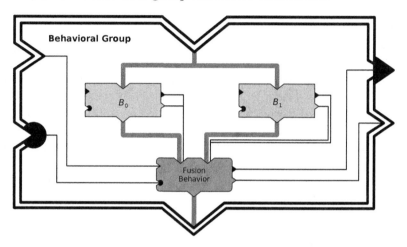

Figure 8.14 Example for a behavioral group in iB2C with a fusion behavior providing the behavior signals for the group interface.

Behavior signal usage The main challenge when coping with systems growing in complexity is making statements about the current system status. This is not only necessary for a developer trying to find out if an implemented feature works, but also for system components trying to reason about the result of a given command. In this sense it becomes invaluable having a common interface of behaviors representing their internal state in an abstract way. In iB2C, behaviors generate the behavior signals *activity* (a) and *target rating* (r) which can be used for detecting several important aspects of the system (for examples see [HBB07]):

- deadlock detection (e. g. by supervising obstacle detection behaviors),

- risk determination (e. g. by supervising slope detection behaviors)

- effort (e. g. by supervising behaviors supervising current measurement of motors)

- oscillation detection (e. g. by supervising behavior activities over time)

8.3.6   Analysis of iB2C networks

Developing robotic systems requires some kind of support for system analysis in order to tackle the complexity of the evolving structure. The aim is to accelerate system development and to reduce time spent for testing. IB2C makes extensive use of the Modular Controller Architecture (MCA). Each behavior is derived from a MCA-module with a standard interface as presented in section 8.3, and with predefined methods for the transfer function and behavior signal calculation. The behaviors are then arranged in a layered network using defined behavior interfaces and interconnections.

Therefore, the behaviors form a graph of interconnected components with a flow of activity which enables analysis using graph theory methods. Figure 8.15 gives an example of an automatically generated iB2C graph containing the flow of activity between behaviors influencing the forward and backward motion of a vehicle. Within this graph, properties like cycles as well as stimulation and inhibition successors and predecessors can be automatically identified to retrieve static information about the influence of behavior modules on the robot's behavior. This way, possible sources of oscillations inside the behavior and interconnections contradicting with the introduced principles have successfully been spotted.

During runtime of the robotic system the software can be supervised by the MCA tools MCAGUI and MCABrowser. The MCAGUI serves as the user interface for the robot which can be configured using predefined widgets and plug-ins. The tool MCABrowser lets the developer have a detailed look at the flow of data during runtime.

For iB2C, the user interface inside MCABrowser is complemented by indicators for the behavior signals. While the robot performs its tasks, a condensed view of the current distribution of activation, activity, and target rating is given using colored bars (see figure 8.16). Additionally, the flow of activity is indicated by colored edges between behavior modules. This way, sources of the current robot behavior can be easily identified and analyzed. The example depicted in figure 8.17 presents three snapshots of the behavior signal visualization. At first the robot is in an idle state where no command is to be executed (top left). Therefore, all behaviors remain inactive. Afterwards, a normal forward motion situation (top right) is indicated by a flow of activity from the top interface to the bottom interface passing through fusion behaviors of different layers. Finally, a situation where the *Forward Tactile Creep* behavior inhibits slow down behaviors while *FW Limit Creep Vel* maintains a minimal creep velocity in order to slowly move into vegetated regions is presented (bottom). This serious situation is clearly visualized by a high target rating of several behaviors as well as the massive

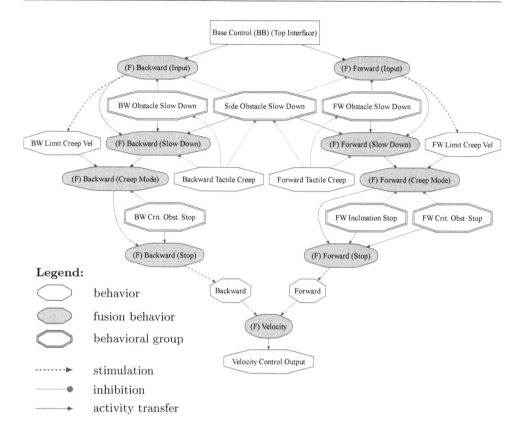

Legend:

- ⬡ behavior
- ⬡ (shaded) fusion behavior
- ⬡ (double) behavioral group
- ┄┄► stimulation
- ───● inhibition
- ───► activity transfer

Figure 8.15 Example for an activity graph of an iB2C behavior network influencing the forward and backward motion of a vehicle. The different styles of the arrows indicate the type of interaction between behaviors, i.e. stimulation, inhibition, or activity transfer. This allows the evaluation of the activity flow through the behavior network.

inhibitory interaction between behaviors. This way, several flaws in implemented iB2C networks have been detected, e. g. behaviors showing no activity due to missing sensor information or errors in the transfer function resulting in contradictory values for activity and target rating.

Figure 8.16 Behavior signal visualization in MCABrowser

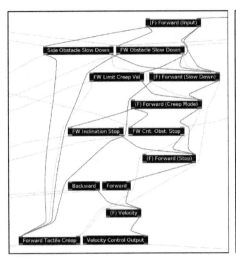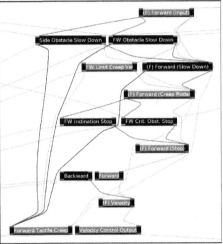

Figure 8.17 Two exemplary snapshots of the on-line behavior signal visualization in MCABrowser (Top: Normal forward motion is indicated by the course of activity through the forward behaviors. Bottom: Obstacles in the robot's proximity result in a high activity of the *Forward Tactile Creep* behavior which inhibits the slow down behaviors in order to move slowly forward)

In cases where it is necessary to guarantee certain system properties, an approach for the formal verification of a subset of behaviors can be followed as described in [PBSS07]. With this method a subset of interconnected behaviors is implemented in the synchronous language Quartz and verified concerning given properties using model checking techniques. Afterwards, code is generated which is periodically called inside a MCA module and which is proven to meet given specifications.

9 Software frameworks

Developing software for autonomous robots from scratch is a complex, time-consuming and error-prone task. There are many issues that need to be dealt with, including hardware access, modeling of the environment, behavior synthesis as well as providing convenient debugging and teleoperation facilities. Especially in larger projects, the software needs to be clearly structured in order to stay maintainable. Ideally, software entities can be easily reused in other projects. Software efficiency and fault-tolerance are further critical aspects.

Therefore, complex robotic applications are typically based on *frameworks*. These frameworks contain common functionality regarding robotic software and completely or partially support many of the issues a programmer would otherwise have to tackle. Hence, a framework has a critical impact on software quality and features of a robot control, as well as the development process in general. It should provide all necessary facilities so that a developer can concisely and conveniently address problems arising in his specific domain.

Many frameworks have been proposed and developed in the past. In fact, "it is only a small overstatement to say that almost every lab has brewed its own solution for robot control architecture, middleware and software integration concepts" [SP07], p. 1. However, not many frameworks are used outside of these labs. Regarding the DARPA Urban Challenge, [MBK07], p. 2, state: "To our knowledge, none of the numerous teams participating in the competition use any of the Robotics Software Systems outside of their original circle of developers".

Different frameworks focus on different aspects such as robustness, efficiency or ease of use. These are supported very well while others are sometimes neglected. Currently, there is no solution which is clearly superior to all the others [MBK07].

Using the vast majority of robotic frameworks, applications are constructed in a modular fashion. Software entities ("modules", "components", "services") are instantiated and can be connected in a network-transparent way. This way, robot controls can be easily distributed across several computing nodes. Therefore, network throughput and latency are critical factors in robotic frameworks.

Other important aspects include ease of use, supported programming languages, generality, flexibility, extensibility, efficiency, scalability, robustness and reliability, real-time capabilities, interoperability, portability and promoting software reuse.[1] Ideally, a framework already provides many hardware drivers and algorithm implementations.

In the following, a small selection of frameworks is presented in more detail. They are used at several labs and are available to the public. First, the Player Project will be introduced. It is arguably the most well-known open source robotics framework. Then, Microsoft Robotics Developer Studio is covered. It provides extensive tool support and was designed to be simple to use. After that, the open source frameworks Orca and MCA are discussed.

The frameworks are presented briefly – with a focus on concepts and general information, since the details will quickly become outdated (major parts are excerpts from [Rei08]).

Other important and interesting frameworks that are not covered in this book include CLARAty [Nes07], ROCI [CCT07], CoolBot [DBHSIGCG07], MARIE [CLR07], Orocos [Bru01], Urbi [Bai07], Robotics4.NET [CCAC07], XCF [FW07], Pyro [BKMY03], ARIA[2] and many more.

9.1 The Player Project

The Player Project[3] is also known as "Player/Stage Project" or "Player/Stage/Gazebo Project". According to [CMG05], the "Player/Stage/Gazebo tools have become a de facto standard in the Open Source robotics community".

Development began in 1999 at the Robotics Research Lab of the University of Southern California[4], "to address an internal need for interfacing and simulation for Multi-Robot Systems" [GVH03], p. 1. In 2001, the project was released under the terms of the GNU General Public License (version 2.1). Since then, it has been downloaded more than 100,000 times and is used in many labs and educational institutions worldwide.[5] In 2006, version 2.0 of the framework was released, a major rework that addressed some of the shortcomings identified in the previous releases. Details concerning these changes can be found in [CMG05]. The framework is still being actively developed.

---

[1] A detailed discussion of these aspects can be found in [Rei08]

[2] http://robots.mobilerobots.com/wiki/ARIA

[3] http://playerstage.sourceforge.net/

[4] http://www.usc.edu/

[5] A list can be found on http://playerstage.sourceforge.net/wiki/PlayerUsers

The Player Project consists of three major parts – "Player", "Stage" and "Gazebo". Central instance of the project is the Player Server. Stage and Gazebo are robot simulators. Stage is a 2D simulator designed to simulate large populations of robots with reasonable accuracy. Gazebo is a 3D simulator with high accuracy and therefore only suited for a small number of robots.

The Player server is a device server that is usually installed on robots, providing access to its sensors and actuators over a network interface. Simply put, the server provides a convenient API to a broad range of commercial robots and robotic hardware from multiple programming languages.

Central design goals were *minimalism* and *simplicity* concerning server and message protocol [GVS+01].

Robot controls are implemented as clients of the Player server. A control can run on the same system as the server, as well as on any other system connected to the robot via network. It may also be distributed across several systems. One design philosophy regarding the player framework was to constrain the development of clients as little as possible and not to impose any architectural or design decisions on them. Furthermore, many *heterogeneous devices* should be concurrently accessible by many *heterogeneous clients* [GVS+01].

Virtually any programming language and any platform with support for TCP sockets can be used for the implementation of clients. However, in practice client-side libraries are used [VGH03] that hide the internals of the player message protocol from the programmer and facilitate development. Such libraries exist for many programming languages including C, C++, Python, Java, LISP, Ruby and Ada. Apart from these libraries, the Player distribution does not contain any facilities or libraries supporting the development of robot controls.

Since Player 2.0, the network interface and protocol are an interchangeable component called "transport". This way, support for technologies like Corba [Obj98] or JINI [Wal99] can be implemented if needed. Currently, a JINI-based transport is available. The original transport mechanism – still the standard mechanism included in the Player distribution – is a clean, simple and efficient protocol based on TCP sockets.

The Player project itself is implemented in C++ and can be compiled and run on many POSIX-compliant platforms including Linux, Solaris, Mac OSX and different variants of BSD. On a Windows host, it may be used inside a Cygwin environment.

Regarding real-time capabilities, the Player project is not suited for meeting hard real-time requirements [CMG05].

The network interfaces that a Player server provides are of particular interest when programming a client. Each hardware device on a robot may provide one or more interfaces for access. These can be existing interfaces, as well as new custom ones.

In the beginning, Player was only supposed to provide simple and flexible interfaces for the "Pioneer" robots used at the University of Southern California [VGH03]. The simulators support the same interfaces, so the control software can be run with either the simulator or the real robots, which is useful for testing. Over time, drivers for further robots were implemented that reused parts of these interfaces. Eventually, the interfaces had significant overlaps. It became apparent that it would be advantageous if robots providing similar functionality implemented the same interfaces. This way, the simulators can be used for different robots. Furthermore, reuse in general is facilitated: The *same* robot control could possibly be run unchanged on *different* robot platforms [VGH03] and would therefore be *device independent* and *portable*. Defining suitable abstract interfaces general enough to support a wide range of similar hardware without becoming too complex is a critical task in this context. Eventually, this led to the *Player Abstract Device Interface* (PADI) that currently contains 42 interfaces for different areas of functionality.[6]

A player server provides a set of such interfaces to access hardware devices. There may be several instances of the same interface type if a robot has multiple sensors of the same type, or other similar subsystems.

The PADI specification is not bound to the Player framework. [VGH03] suggests that a standard robot interface specification like the PADI might be relevant for other robot frameworks as well.

Figure 9.1 (informally) illustrates how a Player server is structured internally. An instance of a Player server manages a set of hardware devices and virtual devices. It has a modular structure so devices can be easily and independently added, exchanged or removed. In order to use a hardware device with the Player server, a driver needs to be implemented that supports the interfaces the device ought to provide. The Player server makes the devices interfaces available to clients. An arbitrary number of clients may connect to a single interface.

---

[6] http://playerstage.sourceforge.net/doc/Player-2.0.0/player/group_
  _interfaces.html

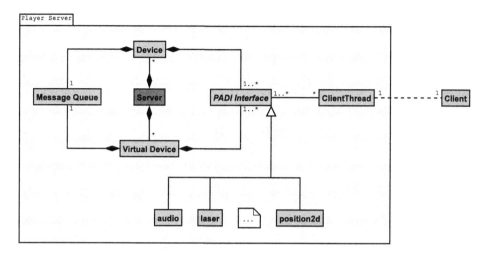

Figure 9.1 Structure of the Player server (from [Rei08])

9.2 Microsoft Robotics Developer Studio

In December 2006, Microsoft released the "Microsoft Robotics Studio" software development kit.[7] The name was changed to "Microsoft Robotics Developer Studio" in 2008. A major design goal was to make developing robotic applications significantly easier than it had been in the past. Developers with little programming experience ("non-programmers" [Mic07]) explicitly belong to the targeted audience. Particularly for them, the Microsoft Visual Programming Language (VPL) was developed, allowing creation of robot controls graphically by connecting services and functional blocks. Apart from that, Microsoft Robotics Developer Studio includes a powerful simulator – the Microsoft Visual Simulation Environment. Microsoft Robotics Developer Studio is currently available in three editions[8] – the "Standard Edition" for $499.95, the "Academic Edition", and the free "Express Edition" with some limitations.

A variety of commercially available robots and other hardware is supported by Robotics Developer Studio. Robotic applications based on Microsoft Robotics Developer Studio require a Microsoft operating system to run. Apart from the desktop operating systems Windows XP, Vista and Server 2003, the operating systems Windows XP Embedded CE 6.0 and Mobile 6.0 are supported. Robot controls are developed using the .NET frame-

---

[7] http://www.microsoft.com/robotics/
[8] Information valid as of June 2009

work and Visual Studio IDEs[9]. Embedded devices require the .NET Compact Framework.

Microsoft Robotics Developer Studio has a strongly *service-oriented* architecture – in some areas identical to web services. A robotic application consists of numerous loosely coupled software components that offer services and are called *Decentralized Software Services* (DSS). These DSS execute inside runtime environments named DSS *Nodes*. The services making up a robot control can be distributed across several nodes running on different systems to form a distributed robot control. Almost anything can be implemented and wrapped as a DSS, including sensors, actuators and different algorithms, but also graphical user interface, or services like "storage".

The DSS infrastructure is implemented on top of the *Concurrency and Coordination Runtime* (CCR). This is a common Microsoft Library that provides sophisticated mechanisms to "manage asynchronous operations, deal with concurrency, exploit parallel hardware and deal with partial failure" [Mic07]. It is used to manage access to shared resources and to implement the loose coupling of DSS through a non-blocking mechanism for sending messages and making remote procedure calls. According to [QFY$^+$07], CCR is implemented in C# and there are plans to develop a C++ version with higher performance.

Important elements of the CCR library are "*Ports* and *PortSet* queuing primitives" as well as *Arbiters*. A Port is a FIFO (First In First Out) queue with items of a specified type [Mic07]. A PortSet is a set of such ports. A DSS, for example, provides a PortSet to receive messages. Arbiters are instances that are assigned to and *observe* a Port. Whenever the Ports' contents change, they are notified and possibly execute user defined code. There are many different types of Arbiters including variants for conditional and nested execution of user code. Apart from that, the CCR contains the *Dispatcher*, *DispatcherQueue* and *Task* classes that are used for scheduling and load balancing [Mic07].

DSS can be implemented in any programming language supported by the Microsoft .NET framework[10] including C#, VB.NET and IronPython. The Microsoft Visual Programming Language (VPL) provided by Robotics Developer Studio may be used as well. Code may also be implemented in unmanaged C++. This is important if performance is critical. Concerning IDEs, any edition of Visual Studio, including the free Express Editions[11], can be used for development. Services can be built either directly using a .NET

---

[9] Integrated Development Environments

[10] http://msdn2.microsoft.com/de-de/netframework/default.aspx

[11] http://www.microsoft.com/express/

IDE or with the MSBuild build tool[12] which is in some respects similar to Apache Ant[13].

Figure 9.2 shows how a DSS is structured.

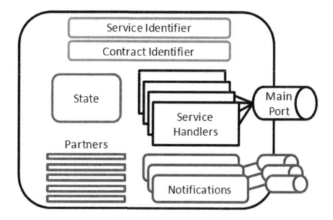

Figure 9.2 Decentralized Software Service structure (from [Mic07])

Each instance of a DSS has a URI as *service identifier*. If a service is, for example, called **exampleservice** and is a singleton, the default URI is:

```
http://host:port/exampleservice
```

If a service may be instantiated more than once, a unique identifier is appended.

In order to be accessible, each service has a *contract*. .NET proxy DLLs are automatically generated from that contract when the binaries of the service are built. These DLLs can be used to access the service. Identifiers of such contracts have the following form [Mic07]:

```
http://schemas.tempuri.org/[year]/[month]/[name].html
```

Furthermore, each service has a *state*. "Any information that is to be retrieved, modified, or monitored as part of a DSS service must be expressed as part of the service state." [Mic07]

Some services require other services for operation. These are the *Partner Services*. A DSS specifies which other services it depends on and which are optional. The DSS Node runtime will attempt to connect the service to its Partner Services at start-up. It can be specified that a service should not start at all if dependencies are missing.

---

[12] http://msdn2.microsoft.com/en-us/library/0k6kkbsd.aspx
[13] http://ant.apache.org/

The *Main Port* is a PortSet (see above) and receives messages from other services. The Main Port must support at least one of these protocols.

For each of the required Ports, a *Service Handler* needs to be implemented. Every time a message is received on that Port, its Service Handler is executed. There is, however, a default implementation of two DSSP operations (`DsspDefaultLookup` and `DsspDefaultDrop`).

A service may subscribe to another service. This way, it will receive a message whenever there is a relevant state change in that other service (*Notifications*). These messages arrive on a separate PortSet. There is one such PortSet for each subscription.

Typically, a robot control application acquires and processes sensor data and operates actuators accordingly. Figure 9.3 illustrates what a typical robot control in Microsoft Robotics Developer Studio looks like:

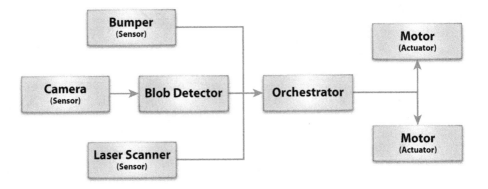

Figure 9.3 Typical structure of a robot control (each box represents a DSS; from [Rei08])

Each sensor and actuator is implemented as a service. Then there are services that depend on other services – the *Blob Detector* in this example. All these services are combined by an *Orchestrator* that ties them together.

As mentioned above, all services are executed inside of a runtime environment referred to as DSS Node. A DSS Node provides facilities for creating, managing and deleting services. This functionality is accessible through a web-based user interface.

Furthermore, running services can be inspected and possibly configured depending on which operations from the HTTP and DSSP protocols they implement. By default, the state of a Service is output in plain XML. Using XSL Transformation Templates, the XML can be transformed to match the look and feel of other running services and the web interface in general.

Regarding security, DSS Nodes provide several mechanisms: NTLM Authentication[14], restricting Service Assembly Loading to Authenticode-signed assemblies, and Network Access Permissions (for details see [Mic07]).

Generally, Microsoft Robotics Developer Studio is not capable of meeting hard real-time requirements since it is based on .NET, and desktop Windows platforms do not have real-time kernels. Regarding this topic, George Chrysanthakopoulos from Microsoft stated, "for hard real time, we recommend a native or even kernel mode component that sits close to the hardware or process that needs strict timing (isochronous), and then communicates filtered, lower frequency data to an MSRS service".[15]

For communication, Microsoft Robotics Developer Studio uses a network protocol referred to as DSSP. DSSP is based on SOAP which is commonly used for web services. SOAP is itself based on XML for data representation and typically HTTP/HTTPS for transport. A short "Specification" of DSSP was released in 2007 (see [NC07]).

DSSP provides commands for creating, inspecting, manipulating and deleting services, as well as subscribing to events that other services produce. This is especially useful for clients to regularly receive updated sensor values. A DSSP connection is established between two services. There are two communication patterns: one-way messages and request/response interactions [NC07].

9.3 Orca

The Orca framework[16] has emerged from parts of the Orocos ("**O**pen **Ro**bot **C**ontrol **S**oftware") project which was funded by the EU. Originally, the Orocos project was planned to be developed cooperatively by three universities in Europe:

- the Royal Institute of Technology in Stockholm (KTH), Sweden[17]

- the Katholieke Universiteit Leuven in Belgium[18]

- the Laboratory for Analysis and Architecture of Systems (LAAS) in Toulouse, France[19]

---

[14] "NT LAN Manager" – an authentication protocol from Microsoft

[15] from http://www.eggheadcafe.com/software/aspnet/29574415/microsoft-robotics-and-re.aspx

[16] http://orca-robotics.sourceforge.net/orca/index.html

[17] http://www.kth.se/

[18] http://www.kuleuven.be/

[19] http://www.laas.fr/

Work started in 2001. Although not an official partner, the FAW in Ulm, Germany, also contributed to the Orocos project.

However, the separate projects never merged and remained largely independent. In 2004, members of the Australian Centre for Field Robotics at the University of Sydney[20] adopted the Swedish part of the Orocos project[21]. Today, they are the maintainers and major contributors of the project. The project was renamed to Orca in 2004 – the part originally developed in Belgium retained the name Orocos[22]. At the end of 2005, Orca 2 was released, addressing some of the shortcomings identified in the previous versions. The most significant change was the switch from CORBA [Obj98] and a custom middleware to a single middleware called ICE[23] ("Internet Communications Engine" – see [HS07] for details).

Orca is an open source project. Large portions of the source code are licensed under the terms of the GNU Lesser General Public License (LGPL) – the remaining parts are available under the GPL, with some minor exceptions.

Orca is mainly developed and used on Linux platforms. However, parts of the framework also run on QNX and Windows operating systems. Furthermore, there are "experimental builds" on Mac OSX platforms.

The central design goal in Orca is promoting software reuse in robotics since this is "the key to making progress in this area" [MBK06], p. 1. Therefore, Orca explicitly follows a *component-based* approach commonly known as Component-Based Software Engineering (CBSE) [BKM+07].

Orca is designed to provide the infrastructure for a functioning component market. Ideally, this would enable using and integrating well-tested third-party components to quickly implement robotic applications with high quality. However, "the benefits of a component-based approach only become apparent when a critical mass of component developers and users arises" [BKM+07], p. 3. According to [BKM+07], the Player Project (see chapter 9.1) is the only robotics framework to have established a significant market.

Orca is meant to be a general purpose framework. It is intended to be suitable for a broad range of robotic systems by imposing "as few constraints as possible" [BKM+07], p. 2 on components and applications. It targets academic as well as commercial applications.

---

[20] http://www.acfr.usyd.edu.au/

[21] http://orca-robotics.sourceforge.net/orca/orca_doc_history.html

[22] http://www.orocos.org/

[23] http://www.zeroc.com/ice.html

Furthermore, *multiple languages* and *multiple platforms* for implementing and running components are supported. Orca is implemented in C++ using "Cross-platform Make"[24] as a building tool.

[BKM+07], p. 2 concisely sums up Orca's architecture: "The design of the Orca framework is conceptually simple. A system consists of a set of components which run asynchronously, communicating with one another over a set of well-defined interfaces. Each component has a set of interfaces it provides and a set of interfaces it requires. The fundamental purpose of the framework is to provide the means for defining and implementing these interfaces. Standardizing the definitions and implementations of interfaces ensures that components are likely to be inter-operable, and hence re-useable."

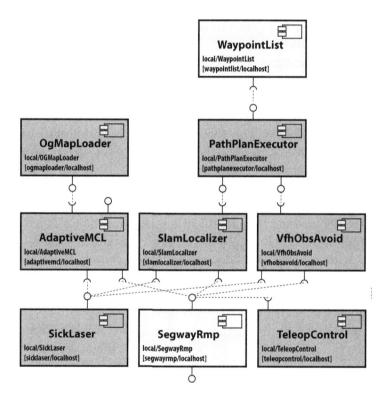

Figure 9.4 Typical Orca-based robot control (here: for Segway robots; source: [BKM+07], p. 15)

The interfaces are specified in "Slice", which is the interface specification language from the ICE middleware. ICE allows using such interfaces in C++,

---

[24] http://www.cmake.org/HTML/Index.html

Java, Python, PHP, C#, Visual Basic and Ruby, so any of these languages can be selected to implement components for the Orca framework.

Apart from the interfaces, Orca imposes no architectural constraints on how the components are implemented. There are guidelines for orientation, however. Orca allows defining new custom interfaces. However, the creation of new interfaces which differ only slightly from existing ones should be avoided. If there were, for example, ten slightly different interfaces for laser scanners, it would be cumbersome for providers of laser scanner drivers to support all of them.

Orca does not provide any mechanisms for real-time communication between components. This choice was taken so as not to impose "unnecessary constraints on those parts of the system which do not require it" [BKM+07], p. 5. If there are real-time requirements, they should be handled within a single component – independent from Orca. The same is true regarding efficiency. Elements requiring maximum efficiency may be implemented in a single Orca component, usually making them less reusable [MBK06], however.

Another aim in Orca was to minimize the code size of Orca's core or "infrastructure" [MBK06]. The switch to the ICE middleware significantly reduced that code size. Currently, the core merely consists of approximately 5500 lines of code.

As mentioned above, a major design change (and arguably improvement) in Orca 2 was the switch from CORBA and a custom middleware to ICE. ICE [HS07] (Internet Communications Engine) is an efficient, object-oriented middleware developed by ZeroC, Inc.[25] and is available both under terms of the GPL as well as under commercial licenses. It is used in many projects around the world. Simply put, the design goal of ICE was to "build a middleware platform that is as powerful as CORBA, without making all of CORBAs mistakes" [HS07], p. 4.

ICE supports several programming languages. Currently, these are C++, C#, Java, Python, Ruby, PHP and Visual Basic. Regarding protocols, TCP as well as UDP can be used. Furthermore, ICE supports features such as data compression or SSL encryption.

Several tools and features provided by ICE are exploited in Orca 2. These include the centralized registry "IceGrid", the application server "IceBox" and the event service "IceStorm".

---

[25] http://www.zeroc.com/

9.4   MCA – Modular Controller Architecture

The development of the MCA framework began at the end of the 1990s in the FZI[26] ("Forschungszentrum Informatik") in Karlsruhe, Germany, when a common platform for all robots at the institute was required [SAG01]. Version two of the framework was released in 2001 under the terms of the GPLv2.[27]

The MCA framework is also used and developed at the Robotics Research Lab at the Technical University of Kaiserslautern. This led to an independent branch that was publicly released in 2007.[28] Having identified some areas for improvement – especially regarding the current networking implementation (see [Rei08]) – a major rework of the framework is almost completed. Java support has already been added (see [KRB08]).

MCA is implemented in C++ and the core of MCA can be compiled on Linux, Windows and Mac OS platforms.

MCA has a *modular* architecture with unified interfaces that enables reuse of many general purpose parts in robot controls. Practice has shown that reuse in MCA actually works well.

Basically, applications based on MCA are divided into "parts", "groups" and "modules". A *part* is compiled into a binary executable and contains groups and modules. A robot control consists of at least one such part. Different parts may be executed on different systems connected via network to create distributed robot controls. This can also be exploited for testing and debugging purposes. Large portions of a robot control can be executed and debugged on an ordinary desktop computer. When everything works satisfactorily, it can simply be moved to the robot completely.

Modules are the basic structural units in MCA. "The concept of MCA is to put basic functional blocks in modules" [SAG01], p. 2. *Groups* in turn contain modules and other groups. They are used to structure modules inside of parts.

Modules, groups and parts have input and output channels that can be connected with *edges* following *data flow* pattern – an output of one entity may be used as an input for the next. Figure 9.5 depicts the central elements of a module in MCA:

A typical module in MCA has two sets of input channels – *SensorInputs* and *ControllerInputs* –, and two sets of output channels – *SensorOutputs* and *ControllerOutputs*. Each channel has a name and a current value. If

---

[26] http://www.fzi.de/ids/
[27] http://www.mca2.org
[28] http://rrlib.cs.uni-kl.de/mca2-kl/

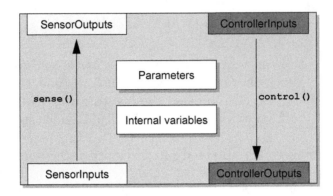

Figure 9.5 Module in MCA (from [Rei08])

two channels are connected with an edge, a value change is automatically forwarded to the destination module. A module has two central methods – `sense()` and `control()` – which are invoked by the framework with a specified rate (usually 1–100 ms). The methods of the different modules are called sequentially in the order of the data flow and this way form a control cycle. There are basically two kinds of data in MCA: sensor data and control data. The first originates from hardware sensors, is subsequently processed by different modules and may reach some kind of user interface. The latter is usually created in some high-level modules or the user interface and finally reaches the actuators of a robot. Obviously, SensorInputs, SensorOutputs and `sense()` deal with sensor data, while ControllerInputs, ControllerOutputs and the `control()` method relate to controller data.

Modules also have *parameters*. These are variables of several simple data types that can be set from the outside. They typically do not change as frequently as data on the sensor and control data paths do. With their help, the behavior of modules can be adapted and optimized at runtime. Finally, modules often have internal variables that, for example, can store a model of the environment or simply the last few sensor values in case of a filter module.

Edges can only be used to forward numbers between modules, which is sometimes a limitation of MCA. The rework, however, will allow any data type. MCA provides a *blackboard* mechanism which can be used to share more complex data between different modules or even parts. Blackboards are areas of shared memory that can be accessed from any module which is part of a robot control. If parts are run on different computers, blackboards can be replicated and synchronized via the network. To avoid conflicts, blackboards may be locked when accessed. Such a lock is mandatory for writing.

One drawback of the current single-buffered blackboard implementation is, however, that threads in robot controls often block waiting for a blackboard lock if the application is not carefully engineered.

There are different types of blackboards which contain all sorts of data such as images, geometry data, text or behavior information. Further types can easily be added.

Figure 9.6 shows an example of how parts and modules are connected on a real robot (screenshots from the MCABrowser tool).

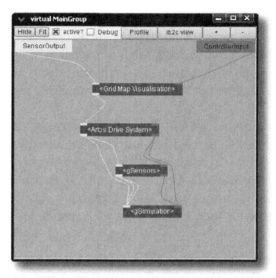

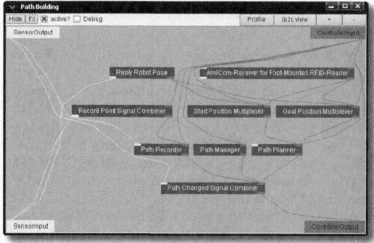

Figure 9.6 Connected MCA parts (top) and MCA modules inside a group (bottom)

Classically, robot controls in MCA have a *hierarchical* structure with groups and modules interacting directly with hardware on the lowest level and more abstract modules on higher levels. Nevertheless, it is also possible to implement robot controls with a flat hierarchy in MCA without problems.

The core of the MCA framework is rather compact, containing only the essential classes for building applications, as well as some tools. It can be extended with MCA libraries that seamlessly integrate into the framework. By now, there is a repository with a large set of libraries that provide modules, groups, hardware support and tool extensions for a wide range of application areas – for example, facilities for mapping, speech synthesis or computer vision. A considerable range of hardware can be accessed using MCA.

MCA was implemented with a focus on efficiency, which is of central importance on robots with tightly limited computing resources. In many places, optimized data structures and shared memory are used to keep the CPU load low.

On platforms running a Real-Time Linux kernel (RTAI and RT-Linux), MCA is generally capable of meeting hard real-time requirements. However, this requires that all the libraries which are used also provide hard real-time guarantees. Many libraries do not. Thus, only parts of the framework are suited for real-time requirements. Apart from that, the blackboard mechanism is only suitable for implementations with real-time requirements when used very carefully.

9.5 Summary and comparison of robotic frameworks

As mentioned in the introduction, there is currently no framework that is clearly superior compared to the others.

Player is popular and easy to use. It provides a convenient API to a wide range of robots. Since "Player seeks to constrain controller design as little as possible" [GVH03], p. 1, however, it hardly provides any facilities for controller design. For complex robotic applications, it is arguably recommendable to use another framework on top of Player as [GCM04] have done. The Player Project is actually used in several other frameworks for accessing hardware including Orca, MARIE and Pyro.

Microsoft Robotics Developer Studio is designed to be easy to use and provides powerful tools. Since it is based on .NET, any programming language supported by the .NET framework can be used to implement robotic applications. However, creating applications with Robotics Developer Studio at some stage requires understanding its service-oriented architecture as well as XML, which is arguably hard for novices – compared to using Player, for instance. [Bai07], p. 5 backs this observation: "Generally speaking, several users have reported that MRS remains relatively complex to master for the moment". Regarding efficiency, the XML-based encoding used in the networking mechanism (DSSP) is not a good choice, since data encoded in XML is much larger than the same data encoded in a binary format. Apart from that, applications developed with Robotics Developer Studio are limited to Windows platforms.

Orca focuses on bringing Component-Based Software Engineering to the Robotics domain. Looking at the author's targets, it appears to be an excellent approach without any major shortcomings. Notably, significant parts of Orca depend on the ICE middleware. Therefore, ICE has a major impact on Orca's performance as well as other critical aspects discussed in the introduction. If maximum efficiency or real-time capabilities are required, the authors recommend implementing all relevant functionality in a single component. There is no support from the framework here.

MCA is one of only very few robotic frameworks with support for hard real-time requirements and it is certainly suitable for large projects. However, the current implementation has a few shortcomings such as the completely synchronous communication mechanism and complex installation and use. A major rework will address these weaknesses and introduce many new features. Experimental versions of this already exist (see [Rei08]).

Frameworks for Robotic Systems

Framework	MCA	Player Project	Microsoft Robotics Developer Studio	Orca
Framework				
(Original) Developer	Forschungszentrum Informatik (FZI), Karlsruhe	Robotics Research Lab, University of Southern California (USC)	Microsoft Corp.	ACFR, University of Sydney, Australia; CAS, Kungliga Tekniska Högskolan, Stockholm;
License	GPL (version 2 or later)	GPL (version 2.1 or later) or LGPL (except of drivers)	Standard Edition (489$), Academic Edition, Express Edition	LGPL and GPL
Homepage	http://www.mca2.org/	http://playerstage.sourceforge.net/	http://www.microsoft.com/robotics/	http://orca-robotics.sourceforge.net/
Development Period	since 1998	since 1999	since 2004	since 2001
First public release	2001	2001	2006	2004 (as "Orca")
Platform				
Operating Systems	Linux, (Windows, Mac OS)	Linux, Solaris, *BSD, Mac OSX	Windows XP, Vista, Server 2003, XP Embedded CE 6.0 and Mobile 6.0	Linux, (QNX, Windows XP, Mac OSX)
Hardware	any suited for above OS – only used on x86; networking requires platforms with same C struct layout	any suited for above OS, including x86, x64, ARM, PPC, Sparc	any suited for above OS, including x86, x64, ARM, MIPS; SuperH	any suited for above OS
General				
Architecture	modular, dataflow-oriented, blackboards	simple, modular, server/client	modular, service-oriented - similar to web services	component-based, minimal core
Structural elements	"parts", "groups", "modules"	"player servers", "clients", "(virtual) devices"	Decentralized System Services (DSS), DSS Nodes	components
Runtime Model	loop-based, one main loop thread per part, thread-containers for further threads*	clients receive data periodically	event-driven	typically loops inside components, at least one thread per component
Interface type	sets of ports or pins with numbers	object-oriented / RPCs, set of standard interfaces	object-oriented / RPCs, set of standard interfaces	object-oriented / RPCs, set of standard interfaces
Real-time capabilities	hard real-time in some parts of the framework	no hard real-time	no real-time inter-service communication	no real-time inter-component communication
Interoperability	(HTTP interface with JavaMCA)*	used in Orca, Marie and Pyro	-	with Player Project (for hardware access)
Development				
Implementation language	C++	C++	limited information, definitely C# and C++, maybe others	C++
Supported languages	C++ (Java with JavaMCA)*	client libraries for C, C++, Python, Java, Ruby, Guile, Octave, LISP, Ada and Matlab	All CLI languages including C++, C#, VB.NET, IronPython, Microsoft Visual Programming Language (VPL), Third party languages	C++, Java, Python, PHP, C#, Visual Basic, Ruby
Build tool	SCons	make	.NET IDE or MSBuild	CMake
Efficient shared memory mechanism for data	yes – however, concurrent access can block*	inside Player server	no	as far as provided by ICE
(SI) units	no explicit conventions, support in some blackboard types*	conventions: distance in metres, angles in radians	no explicit conventions (to our knowledge)	conventions: distance in metres, angles in radians
Communication				
Mechanism exchangeable	no*	yes	no	no
Standard protocol	custom, based on TCP (more than 80 commands)*	custom, simple, based on TCP	DSSP: "simple", based on SOAP; HTTP for web browsers	based on ICE middleware
Data encoding	raw C structs*	XDR	XML	standard ICE encoding
Asynchronous	no*	yes	yes	yes
Pull strategy / push strategy	√/-*	√/√	√/√	√/√
Compression	-*	-	-	√
Automatic resource discovery	no*	no	via UPnP	not really
Web interface	no*	no	yes, for administration & diagnosis	no
Security	password authentication	none	NTLM authentication, policies/permissions	firewall, data encryption
Tools				
GUI Editor	√	?	(.NET IDE)	√
Tool support includes	Debugging/Diagnosis, Simulation (3D)	Debugging/Diagnosis, Simulation (2D & 3D)	Visual Programming Language IDE, advanced Simulation (3D), Testing	Debugging/Diagnosis

* addressed in major rework.

Figure 9.7 Tabular comparison of presented frameworks (June 2009)

Bibliography

[AC02]   P. Althaus and H. I. Christensen. Behaviour coordination
            for navigation in office environments. In *Proceedings of the
            IEEE/RSJ International Conference on Intelligent Robots
            and Systems*, pages 2298–2304, 2002.

[AHB87]   K. S. Arun, T. S. Huang, and S. D. Blostein. Least-squares
            fitting of two 3-d point sets. *IEEE Transactions on Pattern
            Analysis and Machine Intelligence*, 9(5):698–700, Septem-
            ber 1987.

[AKSB07]   C. Armbrust, J. Koch, U. Stocker, and K. Berns. Mo-
            bile robot navigation support in living environments. In
            20. Fachgespräch Autonome Mobile Systeme (AMS), pages
            341–346, Kaiserslautern, Germany, October 2007.

[Alb92]   J.S. Albus. A reference model architecture for intelligent
            systems design. In P. J. Antsaklis and K. M. Passino, edi-
            tors, *An Introduction to Intelligent and Autonomous Con-
            trol*, pages 57–64, Boston, MA, 1992. Kluwer Academic
            Publishers.

[Alb07]   Jan Albiez. *Verhaltensnetzwerke zur adaptiven Steuerung
            biologisch motivierter Laufmaschinen*. GCA Verlag, 2007.

[ALBD03]   J. Albiez, T. Luksch, K. Berns, and R. Dillmann. An
            activation-based behavior control architecture for walking
            machines. *The International Journal on Robotics Research,
            Sage Publications*, vol. 22:pp. 203–211, 2003.

[Ark87]   R.C. Arkin. *Towards Cosmopolitan Robots: Intelligent Nav-
            igation in Extended Man-made Environments*. PhD the-
            sis, Graduate School of the University of Massachusetts,
            September 1987.

[Ark89]   R.C. Arkin. Motor schema-based mobile robot navigation.
            International Journal of Robotics Research, pages 92–112,
            August 1989.

[Ark98]   R. Arkin. *Behaviour-Based Robotics*. MIT Press, 1998.

[Ast07]   X. Astigarraga. 3d reconstruction of structured indoor en-
            vironments. Master's thesis, University of Kaiserslautern,
            March 2007.

[Bai07]   J.-C. Baillie. Design principles for a universal robotic soft-
            ware platform and application to urbi. In *2nd National
            Workshop on Control Architectures of Robots (CAR'07)*,
            pages 150–155, Paris, France, May 31-June 1 2007.

[BDB$^+$07] J. Baeten, K. Donne, S. Boedrij, W. Beckers, and E. Clae-
            sen. Autonomous fruit picking machine: A robotic apple
            harvester. In *6th International Conference on Field and
            Service Robotics - FSR 2007*, 2007.

[Bee96]   R.D. Beer. Toward the evolution of dynamical neural net-
            works for minimally cognitive behavior. *From Animals to
            Animats 4: Proceedings of the Fourth International Confer-
            ence on Simulation of Adaptive Behavior*, 1996.

[BF94]   J. Borenstein and L. Feng. A method for measuring,
            comparing and correcting dead-reckoning errors in mobile
            robots. Technical report, University of Michigan, 1994.

[BILM03]   A. Bonarini, G. Invernizzi, T.H. Labella, and M. Matteucci.
            An architecture to coordinate fuzzy behaviors to control an
            autonomous robot. *Fuzzy Sets and Systems*, 134(1):101–
            115, 2003.

[BKM$^+$07] A. Brooks, T. Kaupp, A. Makarenko, S. Williams, and
            A. Orebäck. Orca: A component model and repository. In
            Brugali [Bru07].

[BKMY03]   D. Blank, D. Kumar, L. Meeden, and H. Yanco. Pyro: A
            python-based versatile programming environment for teach-
            ing robotics. *Journal on Educational Resources in Comput-
            ing (JERIC)*, 3(4), 2003.

[BM92]   P. Besl and N. McKay. A method for registration of 3-
            d shapes. In *IEEE Transactions on Pattern Analysis and
            Machine Intelligence*, volume 14, pages 239–258, February
            1992.

[Bre04]   Christian Brenneke. *Ein scanbasierter Ansatz zur ex-*
             ploratorischen Navigation mobiler Systeme in unstrukturi-
             erten Outdoor-Umgebungen. dissertation.de - Verlag im In-
             ternet GmbH, 2004.

[Bro82]   R. A. Brooks. Solving the find-path problem by represent-
             ing free space as generalized cones. Memorandum MIT-
             Artificial Intelligence Lab, 1982.

[Bro86]   R.A. Brooks. A robust layered control system for a mo-
             bile robot. *IEEE Journal of Robotics and Automation*, RA-
             2(1):14–23, April 1986.

[Bro87]   Rodney A. Brooks. Visual map making for a mobile robot.
             In *Readings in computer vision: issues, problems, principles,*
             and paradigms, pages 438–443. Morgan Kaufmann Publish-
             ers Inc., San Francisco, CA, USA, 1987.

[Bru01]   H. Bruyninckx. Open robot control software: the orocos
             project. In *Proceedings of the IEEE International Con-*
             ference on Robotics and Automation (ICRA), pages 2523–
             2528, Seoul, Korea, May 21-26 2001.

[Bru07]   D. Brugali, editor. *Software Engineering for Experimen-*
             tal Robotics, volume 30 of *Springer Tracts in Advanced*
             Robotics. Springer - Verlag, Berlin / Heidelberg, April 2007.

[Bry01]   J. Bryson. *Intelligence by Design: Principles of Modu-*
             larity and Coordination for Engineering Complex Adaptive
             Agents. PhD thesis, Massachusetts Institute of Technology,
             September 2001.

[CA07]   S. Chernova and R.C. Arkin. From deliberative to rou-
             tine behaviors: a cognitively-inspired action selection mech-
             anism for routine behavior capture. *Adaptive Behavior*
             Journal, 15(2):199–216, 2007.

[CCAC07]   A. Cisternino, D. Colombo, V. Ambriola, and M. Combetto.
             Increasing decoupling in the robotics4.net framework. In
             Brugali [Bru07].

[CCT07]   A. Cowley, L. Chaimowicz, and C. J. Taylor. Roci: Strongly
             typed component interfaces for multi-robot teams program-
             ming. In Brugali [Bru07].

[CLR07]   C. Cote, D. Letourneau, and C. Ra. Using marie for mobile
             robot component development and integration. In Brugali
             [Bru07].

[CMG05]   T. H. J. Collett, B. A. MacDonald, and B. Gerkey. Player
             2.0: Toward a practical robot programming framework.
             In *Australasian Conference on Robotics and Automation
             (ACRA)*, Sydney, Australia, December 5-7 2005.

[CMM06]   Y. Cheng, M.W. Maimone, and L. Matthies. Visual odome-
             try on the mars exploration rovers. *IEEE ROBOTICS AND
             AUTOMATION MAGAZINE*, 13(2):54, 2006.

[CSNP05]   J. Campbell, R. Sukthankar, I. Nourbakhsh, and A. Pahwa.
             A robust visual odometry and precipice detection system
             using consumer-grade monocular vision. In *Proceedings of
             the IEEE International Conference on Robotics and Au-
             tomation*, pages 3421–3427, 2005.

[CSS04]   P. Corke, D. Strelow, and S. Singh. Omnidirectional visual
             odometry for a planetary rover. In *Intelligent Robots and
             Systems (IROS 2004)*, 2004.

[DBHSIGCG07] A. C. Dominguez-Brito, D. Hernandez-Sosa, J. Isern-
             Gonzalez, and J. Cabrera-Gamez. Coolbot: A component
             model and software infrastructure for robotics. In Brugali
             [Bru07].

[Der04]   K.G. Derpanis. The harris corner detector. Technical re-
             port, York University, 2004.

[DH73]   R.O. Duda and P.E. Hart. *Pattern Classification and Scene
             Analysis*. WV, 1973.

[DKWW95]   R. Dillmann, M. Kaiser, F. Wallner, and P. Weckesser. Pri-
             amos: Service, inspection and surveillance tasks. Technical
             report, Institute for Process Control and Robotics, Univer-
             sity of Karlsruhe, 1995.

[Elf89]   A. Elfes. *Occupancy Grids: A Probabilistic Framework for
             Robot Perception and Navigation*. PhD thesis, Department
             of Electrical and Computer Engineering, Carnegie Mellon
             University, Pittsburgh, 1989.

[FM96]   D. Floreano and F. Mondada. Evolution of homing naviga-
           tion in a real mobile robot. *Systems, Man and Cybernetics,
           Part B, IEEE Transactions on*, 26(3):396–407, 1996.

[FN71]   R. Fikes and N.J. Nilsson. Strips: A new approach to the
           application of theorem proving to problem solving. *Artifi-
           cial Intelligence*, 2(3/4):189–208, 1971.

[Fri96]   B. Fritzke. *Handbook of Neural Computation*, chapter Un-
           supervised ontogenic networks. IOP Publishing and Oxford
           University Press, 1996.

[FW07]   J. Fritsch and S. Wrede. An integration framework for de-
           veloping interactive robots. In Brugali [Bru07].

[GB01]   B. Gassmann and K. Berns. Local navigation of lauron iii
           walking in rough terrain. In *CLAWAR 2002, Paris, France*,
           2001.

[GCM04]   J. Giesbrecht, J. Collier, and S. Monckton. Staged experi-
           ments in mobile vehicle autonomy - procedures and results
           with the segwayrmp. Technical report, Defense Research
           and Development Canada - Suffield, December 2004.

[GHvP81]   R. Gerten, R. Hinkel, and E. v. Puttkamer. Coupling two
           microprocessors for the data processing in a micromouse.
           *Implementing Functions; Euromicro 1981; Ed L. Richter,
           P. Le Beux, G. Nagues*, 1981.

[GRD98]   R. Graf, M. Rieder, and R. Dillmann. A nearly holonomous
           driving concept for a mobile robot. Technical report, Insti-
           tute for Process Control and Robotics, University of Karls-
           ruhe, 1998.

[GVH03]   B. Gerkey, R. Vaughan, and A. Howard. The play-
           er/stage project: Tools for multi-robot and distributed sen-
           sor systems. In *11th International Conference on Advanced
           Robotics (ICAR 2003)*, pages 317–323, Coimbra, Portugal,
           June 30 - July 3 2003.

[GVS+01]   B. Gerkey, R. Vaughan, K. Sty, A. Howard, G. Sukhatme,
           and M. Mataric. Most valuable player: A robot device server
           for distributed control. In *Proc. of the IEEE/RSJ Internati-
           nal Conference on Intelligent Robots and Systems (IROS)*,
           pages 1226–1231, Wailea, Hawaii, October 2001.

[HA01]   C. Harper and A.Winfield. Designing behaviour based sys-
              tems using the space-time distance principle. In *Towards
              Intelligent Mobile Robots (TIMR)*, 2001.

[Hac06]   A. Hach. Entwicklung eines positionsbestimmungssystems
              für mobile roboter, optimiert für dsp strukturen. Projektar-
              beit, Technische Universität Kaiserslautern, Januar 2006.

[HBB07]   J. Hirth, T. Braun, and K. Berns. Emotion based con-
              trol architecture for robotics applications. In *Proceedings
              of the Annual German Conference on Artificial Intelligence
              (KI)*, pages 464–467, Osnabrück, Germany, September 10-
              13 2007.

[HK88]   R. Hinkel and T. Knieriemen. Environment perception with
              a laser radar in a fast moving robot. In *IFAC Symposium
              on Robot Control*, pages pp. 68.1–68.7, October 5-7 1988.

[HKNO01]   M. Hashimoto, H. Kawashima, T. Nakagami, and F. Oba.
              Sensor fault detection and identification in dead-reckoning
              system of mobile robot: Interacting multiple model ap-
              proach. Technical report, Department of Mechanical Sys-
              tem Engineering, Hiroshima University, 2001.

[HN68]   P. E. Hart and N. J. Nilsson. A formal basis for the heuristic
              determination of minimum cost paths. *IEEE Transactions
              on System Science and Cybernetics*, 4(2):100 – 107, 1968.

[Hof97]   C. Hofner. *Automatische Kursplanung und
              Fahrzeugführung für mobile Roboter bei flächendeckenden
              Bearbeitungsaugaben.* PhD thesis, Technische Universität
              München, Lehrstuhl für Steuerungs- und Regelungstechnik,
              Prof Dr. Ing G. Schmidt, 1997.

[Hor87]   B. Horn. Closed-form solution of absolute orientation using
              unit quaternions. *Journal of the Optical Society of America*,
              4:629–642, April 1987.

[HS07]   M. Henning and M. Spruiell. *Distributed Programming with
              Ice.* ZeroC, Inc., August 2007.

[Ich05]   D. Ichbiah. *Roboter - Geschichte, Technik, Entwicklung.*
              Knesebeck GmbH und Co Verlags KG, erstausgabe edition,
              2005.

[Joe98]   K. W. Joerg. Mobile robot sonar sensing with pseudo-random codes. In *Proc. IEEE Int. Conf. on Robotics and Automation (ICRA 98) Leuven, Belgium*, 1998.

[KCB97]   J. Kosecka, H. Christensen, and R. Bajcsy. Experiments in behavior composition. *Robotics and autonomous systems*, 19(3-4):287–298, 1997.

[KHB05]   J. Koch, C. Hillenbrand, and K. Berns. Inertial navigation for wheeled robots in outdoor terrain. In *5th IEEE Workshop on Robot Motion and Control (RoMoCo)*, pages 169–174, Dymaczewo, Poland, June 23-25 2005.

[Kos02]   Jana Kostkova. Stereoscopic matching: Problems and solutions, 2002.

[KRB08]   J. Koch, M. Reichardt, and K. Berns. Universal web interfaces for robot control frameworks. In *IEEE/RSJ International Conference on Intelligent Robots and Systems (IROS)*, Nice, France, September 22-26 2008.

[KS97]   K. Konolige and A. Saffiotti. The saphira architecture: A design for autonomy. In *Journal of Experimental and Theoretical Artificial Intelligence (JETAI) 9*, pages pp. 215–235, 1997.

[LDW91]   J.J. Leonard and H. F. Durrant-Whyte. Simultaneous map building and localization for an autonomous mobile robot. In *IEEE/RSJ International Workshop on Intelligent Robots and Systems IROS*, Osaka, Japan, Nov 1991.

[LK81]   B.D. Lucas and T. Kanade. An iterative image registration technique with an application to stereo vision. In *IJCAI81*, pages 674–679, 1981.

[LMD02]   A. Lacaze, K. Murphy, and M. DelGiorno. Autonomous mobility for the demo iii experimental unmanned vehicle, 2002.

[LP79]   T. Lozano-Perez. An algorithm for planning collision-free paths among polyhedral obstacles. *Communications ACM*, 22(10):560 –570, October 1979 1979.

[LRM94]   D. Langer, J. Rosenblatt, and M.Hebert. A behavior-based
            system for off-road navigation. In *IEEE Journal of Robotics
            and Automation*, 1994.

[Lum87]   Vladimir J. Lumelsky. Path-planning strategies for a point
            mobile automaton moving amidst unknown obstacles of ar-
            bitrary shape. *Algorithmica*, 2(1):403–430, 1987.

[Mae89]   P. Maes. The dynamics of action selection. In *IJCAI*, pages
            991–997, 1989.

[Mat89]   L.H. Matthies. *Dynamic stereo vision*. PhD thesis, Carnegie
            Mellon University Pittsburgh, PA, USA, 1989.

[Mat92]   M. Mataric. Integration of representation into goal-driven
            behavior-based robots. *IEEE Transactions on Robotics and
            Automation*, pages 304–312, June 1992.

[Mat97]   Maja J. Matarić. Behavior-based control: Examples from
            navigation, learning, and group behavior. *Journal of Exper-
            imental and Theoretical Artificial Intelligence*, Special issue
            on Software Architectures for Physical Agents, 9(2-3):323–
            336, 1997.

[MBK06]   A. Makarenko, A. Brooks, and T. Kaupp. Orca: Compo-
            nents for robotics. In *IEEE/RSJ International Conference
            on Intelligent Robots and Systems (IROS 2006)*, Beijing,
            China, October 9-15 2006.

[MBK07]   A. Makarenko, A. Brooks, and T. Kaupp. On the benefits
            of making robotic software frameworks thin. In *IEEE/RSJ
            International Conference on Intelligent Robots and Sys-
            tems (IROS 2007)*, San Diego, California, USA, October
            29-November 2 2007.

[Mic07]   Microsoft Corp. *Microsoft Robotics Studio User Guide*,
            2007.

[MMN55]   J. McCarthy, ML Minsky, and N. Nets. A proposal for
            the dartmouth summer research project on artificial intel-
            ligence, 1955.

[Mor96]   Hans Moravec. Robot spatial perception by stereoscopic
            vision and 3d evidence grids. Technical Report CMU-RI-
            TR-96-34, Robotics Institute, Carnegie Mellon University,
            Pittsburgh, PA, September 1996.

[MTKW02]   Michael Montemerlo, Sebastian Thrun, Daphne Koller, and Ben Wegbreit. Fastslam: A factored solution to the simultaneous localization and mapping problem. In *Proceedings of the 18th National Conference on Artificial Intelligence and 14th Conference on Innovative Applications of Artificial Intelligence*, pages 593–598, aug 2002.

[NC07]   H. F. Nielsen and G. Chrysanthakopoulos. *Decentralized Software Services Protocol - DSSP/1.0*. Microsoft Corp., July 2007.

[NDW99]   E. Nebot and H. Durrant-Whyte. Initial calibration and alignment of low cost inertial navigation units for land vehicle applications. Technical report, Department of Mechanical an Mechatronic Engineering, University of Sidney, 1999.

[Nes07]   I. A. Nesnas. The claraty project: Coping with hardware and software heterogeneity. In Brugali [Bru07].

[NM00]   M.N. Nicolescu and M.J. Mataric. Extending behavior-based systems capabilities using an abstract behavior representation. In *Working Notes of the AAAI Fall Symposium on Parallel Cognition*, pages 27–34, North Falmouth, MA, November 3–5 2000.

[NM02]   M.N. Nicolescu and M.J. Mataric. A hierarchical architecture for behavior-based robots. In *Proceedings of the First International Joint Conference on Autonomous Agents and Multi-Agent Systems*, pages 227–233, Bologna, Italy, July 15–19 2002.

[NM03]   M.N. Nicolescu and M.J. Mataric. Linking perception and action in a control architecture for human-robot domains. In *Proceedings of the 36th Annual Hawaii International Conference on System Sciences (HICSS-36)*. IEEE Computer Society, January 6–9 2003.

[Obj98]   Object Management Group, Inc., Framingham, Massachusetts, USA. *The Common Object Request Broker: Architecture and Specification – Version 2.2*, July 1998.

[PBSS07]   M. Proetzsch, K. Berns, T. Schuele, and K. Schneider. For-
             mal verification of safety behaviours of the outdoor robot
             ravon. In *Fourth International Conference on Informatics
             in Control, Automation and Robotics (ICINCO), Angers,
             France*, pages 157–164. INSTICC Press, May 2007.

[PHTO⁺00]   P. Pirjanian, T. Huntsberger, A. Trebi-Ollennu, H. Aghaz-
             arian, H. Das, S. Joshi, and P. Schenker. Campout: a
             control architecture for multirobot planetary outposts. In
             *Proc. SPIE Conf. Sensor Fusion and Decentralized Control
             in Robotic Systems III*, November 2000.

[Pir99]   P. Pirjanian. Behaviour coordination mechanisms — state-
             of-the-art. Technical Report IRIS-99-375, Institute for
             Robotics and Intelligent Systems, School of Engineering,
             University of Southern California, October 7 1999.

[PKEv00]   F. Peters, M. Kasper, M. Eßling, and E. v.Puttkamer.
             Flächendeckendes explorieren und navigieren in a pri-
             ori unbekannter umgebung mit low-cost robotern. vol-
             ume 16. Fachgespräch Autonome Mobile Systeme (AMS
             2000), Karlsruhe, Germany, November 2000. Springer-
             Verlag, Reihe.

[PLB05]   M. Proetzsch, T. Luksch, and K. Berns. Fault-tolerant
             behavior-based motion control for offroad navigation. In
             *20th IEEE International Conference on Robotics and Au-
             tomation (ICRA)*, pages 4697–4702, Barcelona, Spain, April
             18-22 2005.

[PLB10]   M. Proetzsch, T. Luksch, and K. Berns. Development of
             complex robotic systems using the behavior-based control
             architecture ib2c. *Robotics and Autonomous Systems*, 58(1),
             2010.

[PRK91]   D. W. Payton, J. K. Rosenblatt, and D. M. Keirsey. Plan
             guided reaction. In S. S. Iyengar and A. Elfes, editors, *Au-
             tonomous Mobile Robots: Control, Planning, and Architec-
             ture (Vol. 2)*, pages 184–196. IEEE Computer Society Press,
             Los Alamitos, CA, 1991.

[PS03]   R. Philippsen and R. Siegwart. Smooth and efficient ob-
             stacle avoidance for a tour guide robot. In *ICRA*, pages
             446–451, 2003.

[QFY+07] X. Qiu, G. Fox, H. Yuan, S.-H. Bae, G. Chrysanthakopoulos, and H. F. Nielsen. High performance multi-paradigm messaging runtime integrating grids and multicore systems. In *Proceedings of the 3rd IEEE International Conference on e-Science and Grid Computing (eScience 2007)*, Bangalore, India, December 10-13 2007.

[QK93] S. Quinlan and O. Khatib. Elastic bands: Connecting path planning and control. In *Proceedings of IEEE Int. Conference on Robotics and Automation*, pages 802–807, Atlanta, 1993.

[Rei08] M. Reichardt. An advanced framework for robotics. Diploma thesis, Robotics Research Lab - University of Kaiserslautern, Mai 5 2008. unpublished.

[Ros97] J. Rosenblatt. Utility fusion: Map-based planning in a behavior-based system. In *Proceedings of FSR '97 International Conference on Field and Service Robotics*, 1997.

[Sac74] E.D. Sacerdoti. Planning in a hierarchy of abstraction spaces. *Artificial Intelligence*, 5(2):115–135, 1974.

[Sac75] E.D. Sacerdoti. The nonlinear nature of plans. *IJCAI*, 1975.

[SAG01] K. U. Scholl, J. Albiez, and G. Gassmann. Mca- an expandable modular controller architecture. In *3rd Real-Time Linux Workshop*, Milano, Italy, 2001.

[SDC05] M.F. Selekwa, D.D. Dunlap, and E.G. Collins. Implementation of multi-valued fuzzy behavior control for robot navigation in cluttered environments. In *IEEE International Conference on Robotics and Automation, ICRA*, pages 3688–3695, April 2005.

[SHW04] R. D. Schraft, M. Hägele, and K. Wegener. *Service Roboter Visionen*. Hanser Fachbuchverlag, 2004.

[SKBD01] K.-U. Scholl, V. Kepplin, K. Berns, and R. Dillmann. Autonomous sewer inspection: Sensorbased navigation. In *FSR 2001, Int. Conference on Field and Service Robotics*, 2001.

[SKR95] A. Saffiotti, K. Konolige, and E.H. Ruspini. A multivalued logic approach to integrating planning and control. *Artificial Intelligence*, 76(1-2):481–526, 1995.

[SN04]   R. Siegwart and I. Nourbakhsh. *Introduction to Autonomous Mobile Robots*. The MIT Press, 2004.

[SP07]   A. Shakhimardanov and E. Prassler. Comparative evaluation of robotic software integration systems: A case study. In *IEEE/RSJ International Conference on Intelligent Robots and Systems (IROS 2007)*, San Diego, CA, USA, October 29-November 2 2007.

[SPCB06]   J. Shang, T. P.Sattar, S. Chen, and B. Bridge. Design of a climbing robot for inspecting aircraft wings and fuselage. In *Climbing and Walking Robots*, Brussels, Belgium, September 2006.

[SS00]   R. Schraft and G. Schmierer. *Servive Robots*. A K Peters, 2000.

[SS02]   D. Scharstein and R. Szeliski. A taxonomy and evaluation of dense two-frame stereo correspondence algorithms. *International Journal of Computer Vision*, 47(1):7–42, 2002.

[SS04]   R. D. Schraft and F. Simons. Concept of a miniature window cleaning robot - development potentialities for a mass product. In *ISR 2004: 35th International Symposium on Robotics. Proceedings.*, Paris, France, March 2004. International Federation of Robotics.

[SSC90]   R. Smith, M. Self, and P. Cheeseman. Estimating uncertain spatial relationships in robotics. *Autonomous robot vehicles*, pages 167–193, 1990.

[Ste94]   L. Steels. A case study in the behavior-oriented design of autonomous agents. *From Animals to Animats 3: Proceedings of the Third International Conference on Simulation of Adaptive Behavior*, 1994.

[SW95]   A. Schweikard and H. R. Wilson. Assembly sequences for polyhedra. *Algorithmica*, 13:539 – 552, 1995.

[TFBD01]   Sebastian Thrun, Dieter Fox, Wolfram Burgard, and Frank Dellaert. Robust monte carlo localization for mobile robots. *Artificial Intelligence*, 128(1-2):99–141, 2001.

[Thr98]   S. Thrun. Learning metric-topological maps for indoor mo-
             bile robot navigation. In *Artificial Intelligence*, volume 99
             of *1*, pages 21–71, 1998.

[TNS03]   Nicola Tomatis, Illah R. Nourbakhsh, and Roland Siegwart.
             Hybrid simultaneous localization and map building: a nat-
             ural integration of topological and metric. *Robotics and
             Autonomous Systems*, 44(1):3–14, 2003.

[TW97]   David H. Titterton and John L. Weston. *Strapdown inertial
             navigation technology*. Peter Peregrinus Ltd. London, 1997.

[VGH03]   R. Vaughan, B. Gerkey, and A. Howard. On device ab-
             stractions for portable, reusable robot code. In *IEEE/RSJ
             International Conference on Intelligent Robots and Systems
             (IROS 2003)*, pages 2121–2427, Las Vegas, Nevada, USA,
             October 27-31 2003.

[Wal99]   J. Waldo. The Jini architecture for network-centric com-
             puting. *Communications of the ACM*, 42(7):76–82, 1999.

[Web02]   J. Weber. *Globale Selbstlokalisierung für mobile Service-
             Roboter*. PhD thesis, Universität Kaiserslautern, April 2002.

[WGS03]   J. Weingarten, G. Gruener, and R. Siegwart. A fast and
             robust 3d feature extraction algorithm for structured en-
             vironment reconstruction. In *Proceedings of the 11th In-
             ternational Conference on Advanced Robotics (ICAR)*, July
             2003.

[YP05]   Ruigang Yang and Marc Pollefeys. A versatile stereo im-
             plementation on commodity graphics hardware, 2005.

[ZvP94]   U. R. Zimmer and E. von Puttkamer. Realtime-learning
             on an autonomous mobile robot with neural networks. In
             Euromicro - Realtime-Workshop, 1994.

Index

Understanding IT

Eberhard Sturm
The New PL/I
... for PC, Workstation and Mainframe
2009. X, 304 pp. with 80 Fig. and Online Service Softc. EUR 59,90
ISBN 978-3-8348-0726-7

Klaus D. Niemann
From Enterprise Architecture to IT Governance
Elements of Effective IT Management
2006. xii, 232 pp. with 89 figs. and Online-Service. Softc. EUR 61,90
ISBN 978-3-8348-0198-2

Diffenderfer, Paul M.; El-Assal, Samir
Microsoft Dynamics NAV
Jump Start to Optimization
2., rev. Ed. 2008. XII, 304 pp. with 209 fig.
softc. EUR 49,90
ISBN 978-3-8348-0516-4

Karsten Berns | Ewald von Puttkamer
Autonomous Land Vehicles
Steps towards Service Robots
2009. approx. 250 pp. with 100 Fig. softc. approx. EUR 34,90
ISBN 978-3-8348-0421-1

VIEWEG+ TEUBNER
Abraham-Lincoln-Straße 46
65189 Wiesbaden
Fax 0611.7878-400
www.viewegteubner.de

Stand Juli 2009.
Änderungen vorbehalten.
Erhältlich im Buchhandel oder im Verlag.

Grundlagen verstehen und umsetzen

Wolfgang Ertel
Grundkurs Künstliche Intelligenz
Eine praxisorientierte Einführung
2., überarb. Aufl. 2010. XII, 342 S. mit 127 Abb. und Online-Service
(Computational Intelligence) Br. EUR 22,90   ISBN 978-3-8348-0783-0

Peter Mandl | Andreas Bakomenko |Johannes Weiß
Grundkurs Datenkommunikation
TCP/IP-basierte Kommunikation: Grundlagen, Konzepte und Standards
2008. XII, 402 S. mit 219 Abb. und Online-Service Br. EUR 29,90
ISBN 978-3-8348-0517-1

René Steiner
Grundkurs Relationale Datenbanken
Einführung in die Praxis der Datenbankentwicklung für Ausbildung, Studium
und IT-Beruf
7., überarb. u. akt. Aufl. 2009. XI, 235 S. mit 160 Abb. und Online-Service Br.
EUR 24,90   ISBN 978-3-8348-0710-6

Frank Klawonn
Grundkurs Computergrafik mit Java
Die Grundlagen verstehen und einfach umsetzen mit Java 3D
2., erw. Aufl. 2009. XII, 305 S. mit 139 Abb. u. 6 Tab. und Online-Service
Br. EUR 24,90   ISBN 978-3-8348-0691-8

**VIEWEG+
TEUBNER**
Abraham-Lincoln-Straße 46
65189 Wiesbaden
Fax 0611.7878-400
www.viewegteubner.de

Stand Juli 2009.
Änderungen vorbehalten.
Erhältlich im Buchhandel oder im Verlag.

IT-Berufe

Manfred Wünsche

Prüfungsvorbereitung für IT-Berufe

Die wirklich wichtigen Prüfungsinhalte, nach Lernfeldern sortiert -
Übungsaufgaben mit kommentierten Lösungen
4., akt. Aufl. 2009. XVI, 240 S. mit 29 Abb. und Online-Service Br. EUR 19,90
ISBN 978-3-8348-0720-5

Heinz Burnus

Datenbankentwicklung in IT-Berufen

Eine praktisch orientierte Einführung mit MS Access und MySQL
2008. XII, 113 S. mit 123 Abb. und Online-Service
Br. EUR 15,90
ISBN 978-3-8348-0152-4

Markus von Rimscha

Algorithmen kompakt und verständlich

Lösungsstrategien am Computer
2009. VIII, 144 S. mit 60 Abb. u. 35 Tab. und Online-Service Br. EUR 19,90
ISBN 978-3-8348-0569-0

Hans Brandt-Pook | Rainer Kollmeier

Softwareentwicklung kompakt und verständlich

Wie Softwaresysteme entstehen
2008. XVI, 190 S. mit 53 Abb. und Online-Service
Br. EUR 19,90
ISBN 978-3-8348-0365-8

VIEWEG+
TEUBNER

Abraham-Lincoln-Straße 46
65189 Wiesbaden
Fax 0611.7878-400
www.viewegteubner.de

Stand Juli 2009.
Änderungen vorbehalten.
Erhältlich im Buchhandel oder im Verlag.

Printed in the United States
By Bookmasters